RESEARCH AND PERSPECTIVES IN NEUROSCIENCES

Fondation Ipsen

Editor

Yves Christen, Fondation Ipsen, Paris (France).

Editorial Board

Springer
Berlin
Heidelberg
New York
Barcelona
Hong Kong
London
Milan
Paris
Tokyo

C. E. Henderson   D. Green
J. Mariani   Y. Christen (Eds.)

Neuronal Death by Accident or by Design

With 29 Figures and 4 Tables

 Springer

Henderson, Christopher E., Ph. D.
INSERM U.382 – IBDM
Campus de Luminy, Case 907
13288 Marseille Cedex 09, France
E-mail: chris@ibdm.univ-mrs.fr

Green, Douglas R., Ph. D.
La Jolla Institute for Allergy
and Immunology
Division of Cellular Immunology
10355 Science Center Drive
San Diego, CA 92121, USA
E-mail: dgreen5240@aol.com

Mariani, Jean, Ph. D.
CNRS et Université Pierre
et Marie Curie
Institut des Neurosciences, UMR 7624
9 quai Saint Bernard
75005 Paris, France
E-mail: jean.mariani@snv.jussieu.fr

Christen, Yves, Ph. D.
Fondation IPSEN
Pour la Recherche Thérapeutique
24, rue Erlanger
75781 Paris Cedex 16, France
E-mail:
yves.christen@beaufour-ipsen.com

ISBN 3-540-41777-X Springer-Verlag Berlin Heidelberg New York

Library of Congress Cataloging-in-Pulication Data
Neuronal death by accident or by design / C.E. Henderson ... [et al.]. p. cm. – (Research and perspectives in neurosciences). Includes bibliographical references and index.
ISBN 354041777X (alk. paper)
1. Neurons – Congresses. 2. Aptoptosis – Congresses. I. Henderson, C.E. (Christopher E.) II. Series.

Springer-Verlag Berlin Heidelberg New York
a member of BertelsmannSpringer Science+Business Media GmbH

http://www.springer.de

© Springer-Verlag Berlin Heidelberg 2001
Printed in Germany

The use of general descriptive names, registered names, trademarks, etc. in this publication does not imply, even in the absence of a specific statement, that such names are exempt from the relevant protective laws and regulations and therefore free for general use.

Product liability: The publishers cannot guarantee the accuracy of any information about the application of operative techniques and medications contained in this book. In every individual case the user must check such information by consulting the relevant literature.

Production: PRO EDIT GmbH, 69126 Heidelberg, Germany
Cover design: design & production, 69121 Heidelberg, Germany
Typesetting: K+V Fotosatz GmbH, 64743 Beerfelden, Germany

SPIN 10832572   27/3130/göh-5 4 3 2 1 0 – Printed on acid-free paper

Preface

The nervous system was one of the first sites in which the process of cell death was thoroughly studied, and the explosion of research in this area over the last decade has brought with it huge advances in our understanding of the molecular and cellular mechanisms involved. Nevertheless, the excessive neuronal death in the brain that underlies neuro-degenerative disease is still essentially incurable. Alzheimer's disease, Parkinson's disease, Huntington's disease, and amyotrophic lateral sclerosis (ALS) kill tens of thousands of people and yet we still now very little about the pathological processes that lead from the initial trigger to degeneration and death.

The aim of the *Colloque Médecine et Recherche* organized by the Fondation Ipsen in Paris (October 9, 2000) upon which this book is based was to bring together scientists from both ends of the spectrum of research into neuronal cell death. On one hand, four chapters represent the intense current effort to understand the way in which the mitochondrion controls the activation of the final stages of cell death. Such has been the impact of this area of research that it is now almost impossible to remember that, only a few years ago, very few of us would have singled out this organelle as a site for key events in apoptosis, and none of us would have guessed that cytochrome c played an irreplaceable role in the process.

Another four articles attack the problem from the other side. How do specific insults in particular human or mouse neuro-degenerative diseases translate into mechanisms that will not only allow us to better understand what is happening in these patients but also, with luck, allow for development of more efficient and specific drugs in the future? In the middle of the spectrum, developmental neuronal death is demonstrated by four other chapters to be far more complex and varied than one would have supposed even five years ago.

Each chapter provides fascinating new insights into the mechanisms underlying neuronal loss. Taken together, however, they seem to point to three general messages:

First, the concept of a central common cell death pathway, originally derived from studies on the nematode, has been an outstandingly productive paradigm in bringing together different strands of research from fields as different as oncology, developmental biology and neurology. However, the complexity of the nervous system perhaps strains it to the limit. Different neurons die by different mechanisms, and even similar neurons can engage different pathways as a response to different death triggers.

Secondly, truly striking links have been made between results obtained in the culture dish (or even cell-free systems) and the diseased human brain. Nevertheless, there are still major grey areas that differentiate the two types of study. The duration of human pathological neuronal loss over periods of years has attracted much interest and speculation, but as yet no truly convincing explanation. Neurons might for long periods be unhealthy (and therefore dysfunctional) and this state clearly requires more characterization and study.

Lastly, in spite of the thousands of articles in all the best journals, we still have a lot to learn. This is as true for the complex ballet of interactions between Bcl-2 family members at the mitochondrial membrane upstream of cytochrome c release as it is for more general problems such as the means by which cells can engage caspase-independent programmed cell death. In either case, this knowledge will be vital for any structured approach to the problem of neuronal cell death, and probably necessary if the degeneration and loss of neurons in human patients is ever to be prevented for a significant period of time.

We hope that the articles assembled here will provide a useful contribution to ongoing changes in our way of thinking. We take this opportunity of taking Jacqueline Mervaillie and her staff for their outstanding skill and tact in the organization of the meeting and Mary Lynn Gage for editorial assistance.

Summer 2001   Christopher Henderson
                                         Douglas Green
                                         Jean Mariani
                                         Yves Christen

Contents

List of Contributors

BAKOUCHE, J.
Laboratoire Développement et Vieillissement du Système Nerveux,
Institut des Neurosciences, UMR 7624, CNRS et Université Pierre
et Marie Curie, Boite 14, 9 quai Saint Bernard, 75005 Paris, France

CAMPANA, A.
Laboratoire Développement et Vieillissement du Système Nerveux,
Institut des Neurosciences, UMR 7624, CNRS et Université Pierre
et Marie Curie, Boite 14, 9 quai Saint Bernard, 75005 Paris, France

DA CRUZ, S.
Département de Biologie Cellulaire, Sciences III, 30, quai E-Ansermet,
1211 Genève 4, Switzerland

DESHMUKH, M.
Departments of Neurology and Molecular Biology and Pharmacology,
Washington University School of Medicine, Saint Louis, MO 63110, USA

ESTEVEZ, A.
Department of Physiology and Biophysics, University of Alabama,
Birmingham, AL 35294, USA

ETHELL, D. W.
Division of Cellular Immunology, La Jolla Institute for Allergy
and Immunology, 10355 Science Center Drive, San Diego, CA 92121, USA

FLAVELL, R. A.
Howard Hughes Medical Institute, Section of Immunobiology,
Yale University School of Medicine, New Haven, CT 06510, USA

GREEN, D. R.
Division of Cellular Immunology, La Jolla Institute for Allergy
and Immunology, 10355 Science Center Drive, San Diego, CA 92121, USA

HENDERSON, C. E.
INSERM U.382, IBDM (CNRS – INSERM – Univ. Méditerranée),
Campus de Luminy – Case 907, 13288 Marseille Cedex 9, France

HIRSCH, E. C.
INSERM U289, Neurologie et Thérapeutique Expérimentale,
Hôpital de la Salpêtrière, 47, Boulevard de l'Hôpital, 75651 Paris Cedex 13,
France

HUMBERT, S.
Institut Curie, Laboratoire Raymond LATARJET, Bat. 110–112,
Centre Universitaire, 91405 Orsay Cedex, France

IKONOMIDOU, C.
Department of Pediatric Neurology, Charité, Virchow Clinics,
Humboldt University, Augustenburger Platz 1, 13353 Berlin, Germany

JOHNSON, E. M. JR.
Departments of Neurology and Molecular Biology and Pharmacology,
Washington University School of Medicine, Saint Louis, MO 63110, USA

KROEMER, G.
Centre National de la Recherche Scientifique, UMR 1599,
Institut Gustave Roussy, 39 rue Camille-Desmoulins, 94805 Villejuif, France

KUAN, C.-Y.
Section of Neurobiology, Yale University School of Medicine, C310A SHM,
333 Cedar Street, New Haven, CT 06510, USA

LOHOF, A.
Laboratoire Développement et Vieillissement du Système Nerveux,
Institut des Neurosciences, UMR 7624, CNRS et Université Pierre
et Marie Curie, Boite 14, 9 quai Saint Bernard, 75005 Paris, France

MARIANI, J.
Laboratoire Développement et Vieillissement du Système Nerveux,
Institut des Neurosciences, UMR 7624, CNRS et Université Pierre
et Marie Curie, Boite 14, 9 quai Saint Bernard, 75005 Paris, France

MARTINOU, J.-C.
Département de Biologie Cellulaire, Sciences III, 30, quai E-Ansermet,
1211 Genève 4, Switzerland

OPPENHEIM, R. W.
Department of Neurobiology and Anatomy and the Neuroscience Program,
Wake Forest University Medical School, Winston-Salem, NC 27157, USA

PETTMANN, B.
INSERM U.382, IBDM (CNRS – INSERM – Univ. Méditerranée),
Campus de Luminy – Case 907, 13288 Marseille Cedex 9, France

PREVETTE, D.
Department of Neurobiology and Anatomy and the Neuroscience Program,
Wake Forest University Medical School, Winston-Salem, NC 27157, USA

PUTCHA, G. V.
Departments of Neurology and Molecular Biology and Pharmacology,
Washington University School of Medicine, Saint Louis, MO 63110, USA

RAKIC, P.
Section of Neurobiology, Yale University School of Medicine, C310A SHM,
333 Cedar Street, New Haven, CT 06510, USA

RAOUL, C.
INSERM U.382, IBDM (CNRS – INSERM – Univ. Méditerranée),
Campus de Luminy – Case 907, 13288 Marseille Cedex 9, France

RAVAGNAN, L.
Centre National de la Recherche Scientifique, UMR 1599,
Institut Gustave Roussy, 39 rue Camille-Desmoulins, 94805 Villejuif, France

ROUCOU, X.
Département de Biologie Cellulaire, Sciences III, 30, quai E-Ansermet,
1211 Genève 4, Switzerland

SANCHEZ, B.
Département de Biologie Cellulaire, Sciences III, 30, quai E-Ansermet,
1211 Genève 4, Switzerland

SAUDOU, F.
Institut Curie – Research Division, UMR 146 CNRS, Bat. 110–112,
Centre Universitaire, 91405 Orsay Cedex, France

SELIMI, F.
Laboratoire Développement et Vieillissement du Système Nerveux,
Institut des Neurosciences, UMR 7624, CNRS et Université Pierre
et Marie Curie, Boite 14, 9 quai Saint Bernard, 75005 Paris, France

TERRADILLOS, O.
Département de Biologie Cellulaire, Sciences III, 30, quai E-Ansermet,
1211 Genève 4, Switzerland

UGOLINI, G.
INSERM U.382, IBDM (CNRS – INSERM – Univ. Méditerranée),
Campus de Luminy – Case 907, 13288 Marseille Cedex 9, France

VOGEL, M. W.
Maryland Psychiatric Research Center, University of Maryland Medical
School, P.O. Box 21247, Baltimore, MD 21228, USA

YAGINUMA, H.
Department of Anatomy, School of Medicine, Fukushima Medical University,
Fukushima 960-1295, Japan

Mitochondria and Apoptosis, the Stepping Stones on the Path to Death

D. W. Ethell and D. R. Green

Summary

Apoptosis is a form of cellular suicide that allows for the removal of damaged, infected, superfluous or otherwise questionable cells without releasing toxic cellular contents that may trigger an inflammatory response, or damage nearby cells. The active and orderly characteristics of apoptosis contrast with necrotic cell death, which is stochastic and usually pro-inflammatory. The biochemical pathways involved in apoptosis involve the activation of a family of constitutively expressed proteinases called caspases. Working backwards from the last stages of cell death toward caspase activation we can now determine which apoptotic pathways are engaged in response to particular pro-apoptotic stimuli. Among the beneficiaries of this newfound knowledge are those investigators working on neurodegenerative disorders that result from the apoptotic death of discrete neuronal populations. Applying current knowledge of the mechanisms of apoptosis to long-studied disorders such as Huntington's, Parkinson's and Alzheimer's diseases is a bit like turning to the back of a mystery novel halfway through. If the middle chapters of that book are missing, knowing how it ends will aid greatly in reconstructing what happened, despite taking out some of the mystery. The detailed characterization of apoptosis is providing an end-point from which we can trace these disorders backwards to their causes. Insights provided by apoptosis research will speed progress and aid in the development of efficacious treatments that address the causes, rather than just the symptoms, of many neurodegenerative disorders.

Apoptotic Execution

All metazoan cells contain the necessary biochemical machinery for their own apoptotic demise. Preventing the activation of this machinery seems to be an ongoing process and those cells not up to the task are quickly eliminated (Weil et al. 1996). This Darwinian approach to cellular survival is critical in the development and homeostasis of all multicellular animals. Too little cell death and the organism runs the risk of neoplasia; alternatively, too much apoptosis may prove equally disastrous. Knockout studies of key apoptosis genes have regularly shown devastating phenotypes, indicating their im-

Henderson/Green/Mariani/Christen (Eds.)
Neuronal Death by Accident or by Design
© Springer-Verlag Berlin Heidelberg 2001

portance in development. Indeed, given the significance of apoptotic regulation to the trillions of cells in the human body, it is not surprising that life and death decisions are so tightly regulated.

Apoptotic cell death is classically defined by the morphologic appearance of the dying cell, and this remains so despite our current understanding of the biochemistry of the process. The cell undergoing apoptosis may display some or all of the following features: blebbing, chromatin condensation, nuclear fragmentation, loss of adhesion and rounding (in adherent cells), and cell shrinkage. Often the DNA is cleaved, and this fragmentation can be seen as an oligonucleosomal "ladder" when separated by gel electrophoresis. It is important to note that ladder formation or its absence, per se, is not proof that apoptosis has or has not occurred; bona fide apoptosis can occur in the absence of a DNA ladder, and laddering can occur in necrotic cells (Collins et al. 1992). The TUNEL assay, which detects double strand DNA breaks in cells, should also be evaluated with caution, as it can label both apoptotic and necrotic cells.

During apoptosis, cells alter their plasma membrane so as to be rapidly taken up by both "professional" (macrophages, dendritic cells, microglia) and "nonprofessional" (e.g., epithelial cells) phagocytes. While several receptors on phagocytes have been implicated in this clearance process (Savill et al. 1990; Ren et al. 1995; Fadok et al. 1998; Devitt et al. 1998), only one change in the apoptotic cell has been identified (although others are suspected), the externalization of phosphatidylserine (PS). A receptor for PS on phagocytes has recently been identified as a major player in the recognition of apoptotic cells (Fadok et al. 2000). PS is normally found sequestered to the inner leaflet of the plasma membrane, and during apoptosis it equilibrates to both inner and outer leaflets. This externalization can be most easily detected by the use of annexin V-FITC, which specifically binds PS (Martin et al. 1995), but care must be taken in interpreting results, since any disruption of the plasma membrane will expose PS.

In most cases, the apoptotic phenomena described above are a result of the cleavage of specific substrates by caspases, a family of cysteinyl-dependent aspartate-specific proteinases that become active during apoptosis (Alnemri et al. 1996). Several different caspases are constitutively present in cells, and these reside in the cytosol in an inactive "zymogen" form (Wolf et al. 1999). The most prevalent caspase is caspase-3, which is ultimately responsible for the majority of the apoptotic effects, although it is supported by two others, caspases-6 and -7. These three caspases are often referred to as the "executioner caspases" to indicate their role in coordinating the death of the cell. However, when the executioner caspases become active, during apoptosis, they do not themselves kill the cell (i.e., these are not digestive enzymes). It is their cleavage of specific substrates that results in the changes we see and describe as apoptosis.

High and low molecular weight DNA fragmentation is caused by the action of caspase-3 on a nuclease (CAD or DFF45)/inhibitor (iCAD or DFF40) complex (Enari et al. 1998; Liu et al. 1997). Caspase-3 cleaves the inhibitor, allowing the nuclease to cut the chromatin (it appears that the amount of activated nuclease determines whether or not DNA fragmentation will proceed to a ladder; Fig. 2). Thymocytes deficient for CAD can undergo apoptosis both in vitro and in vivo, though there is not DNA laddering (Zhang et al. 2000). Following DNA cleavage, the chromatin does condense, but this may be due to another caspase substrate, called acinus (Sahara et al. 1999).

Blebbing is orchestrated via the cleavage and resulting activation of gelsolin (Kothakota et al. 1997) and p21-activated kinase-2 (Lee et al. 1997; Rudel and Bokoch 1997), and probably through cleavage of fodrin (Martin et al. 1995) to dissociate the plasma membrane from the cytoskeleton. The result is probably an effect of microfilament tension and local release, since the depolymerization of actin prevents blebbing (Cotter et al. 1992). Interestingly, in some cells undergoing apoptosis, blebbing appears to occur independently of caspase activity (McCarthy et al. 1997), but in others it is caspase dependent.

The externalization of PS during apoptosis is usually caspase dependent (Martin et al. 1996), although the precise mechanisms have not been elucidated. In some cells, PS externalization appears to be caspase independent (Vanags et al. 1996) although, again, the mechanism is unknown. For example, PS externalization in platelets does not involve caspases (Wolf et al. 1999).

Caspases

The ability of caspases to activate other caspases leads to the idea of a caspase cascade that eventually activates executioner caspases, which can activate all remaining caspases as well as cleaving key substrates. However, getting the ball rolling, so to speak, requires the cleavage and activation of apical, or "initiator", caspases. Activation of initiator caspases is accomplished in association with other proteins in complexes called apoptosomes. In our discussions, we will focus on two kinds of apoptosomes that activate the initiators, caspase-8 and caspase-9, which have very different functions in triggering apoptosis. Both of these have in common an excellent ability to cleave and activate executioner caspases, although their regulation is disparate.

A major difference between initiator and executioner caspases is the nature of their prodomains. Executioner caspases have very short prodomains, whereas those of the initiators are longer and contain special protein interaction motifs (Wolf and Green 1999). These interaction motifs bind to adapter molecules that permit the autoactivation of the initiator caspases. For the two initiator caspases discussed here this happens in different ways (Fig. 1). Caspase-8 interacts with its adapater molecule via a motif called the death effector domain (DED), whereas caspase-9 interacts with a different adapter molecule via a caspase recruitment domain (CARD) (Thornberry and Lazebnik 1998; Green 1998). Interestingly, the protein interaction motifs in these

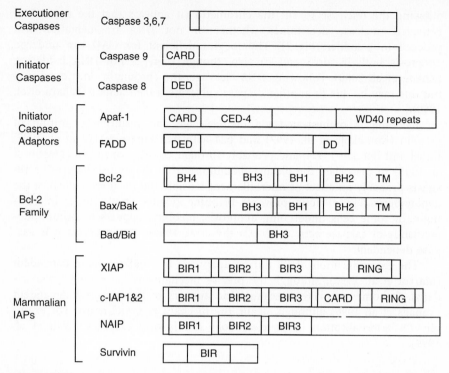

Fig. 1. Schematic alignments of proteins representative of those involved in apoptosis. Excutioner caspases-3,-6,-7 have very small prodomains (indicated by a small box) compared with initiator caspases, which contain speciaslized amino-terminal protein-protein interaction motifs (caspase recruitment domain, CARD, or death effector domain, DED) used for binding to the initiator caspase adaptor proteins shown directly below. Apaf-1 also has a carboxyl-terminal WD40 repeat region. Representative members of the Bcl-2 family are shown with the Bcl-2 homology regions indicated by BH1-4. Both anti-apoptotic Bcl-2 and pro-apoptotic Bax/Bak proteins contain a transmembrane domain near the carboxyl-terminus. Mammalian members of the IAP family are shown, with the baculoviral IAP region domains indicated (BIR1-3). XIAP and cIAP1 & 2 have Zn^{2+} ring finger domains near their carboxyl-termini. Although there are other mammalian BIR-containing proteins in this group, those have not been demonstrated to inhibit apoptosis. This list is not exhaustive and concerns only mammalian homologs. The relative sizes of the proteins and domains are not to scale

two caspases are structurally very similar, but unrelated at the level of primary sequence. This characteristic is also shared by another protein interaction domain important in apoptosis, the death domain (DD), whose function will be discussed below (DDs are not present in caspase prodomains).

Apoptosomes

The first kind of apoptosome we will consider forms on the cytoplasmic tails of ligand-bound death receptors. Perhaps the simplest and most well characterized death receptor is Fas. Extracellular binding of trimeric Fas ligand (FasL) causes activation of Fas (Fig. 2). Death domains on the carboxyl-termini of Fas are then able to recruit the adapter protein, FADD, which also contains a death domain. FADD also contains another protein interaction motif, the death effector domain (DED), which in turn recruits another DED-containing protein, the initiator procaspase-8. The trimeric nature of Fas oligomerization allows for the recruitment of multiple FADD and procaspase-8 proteins into the apoptosome. Zymogenic procaspases have very little cleavage activity, (approximately 1% of maximal activity), but the close proximity of several procaspase-8 molecules in this apoptosome facilitates cross-activation. Active caspase-8 is then able to cleave and activate other caspases, including caspase-3.

Signaling between FasL and Fas was initially thought to happen only between cells (trans). However, it is clear that some Fas-expressing cells can undergo apoptosis if they begin to express FasL (cis), even at limiting dilutions. Thus factors influencing the upregulation of FasL may be pro-apoptotic to cells already expressing Fas. There are potential consequences for this finding within the nervous system, as many cells express low levels of Fas. Cerebellar granular neurons are supported in medium containing 25 mM KCl, which is sufficient to depolarize the cells. However, decreasing extracellular KCl to 5 mM stresses the cells, which begin to express FasL and subsequently become apoptotic (LeNiculescu et al. 1998). This effect may also work within the context of Aβ-mediated neurotoxicity seen in Alzheimer's disease. We have recently found that Aβ-mediated neuronal apoptosis is blocked by FasFc fragments in vitro. Further, neurons isolated from mice with mutations in Fas[lpr] or FasL[gld] are resistant to Aβ-mediated apoptosis, suggesting that this pathway is involved (Ethell et al. in preparation).

Mitochondria-Dependent Apoptosome

A second kind of apoptosome that we refer to as mitochondria-dependent apoptosomes (MDA) are dependent upon factors released from mitochondria. Formation of MDA involves a metameric cytoplasmic structure consisting of procaspase-9 (the initiator caspase), Apaf-1, cytochrome c and dATP/ATP as an essential co-factor (Zou et al. 1997). In purified cell-free systems, the addition of these four components is necessary and sufficient for high caspase-9 activity (Stennicke and Salvesen 1999). Caspase-9 activity does not strictly depend on cleavage as even cleaved caspase-9 is inactive unless bound by Apaf-1, which in turn hydrolyzes dATP (or ATP) at two conserved p-loops. The necessity of cytochrome c for Apaf-1 to efficiently bind caspase-9 is clear, yet its precise role in the apoptosome is enigmatic. Currently, it is

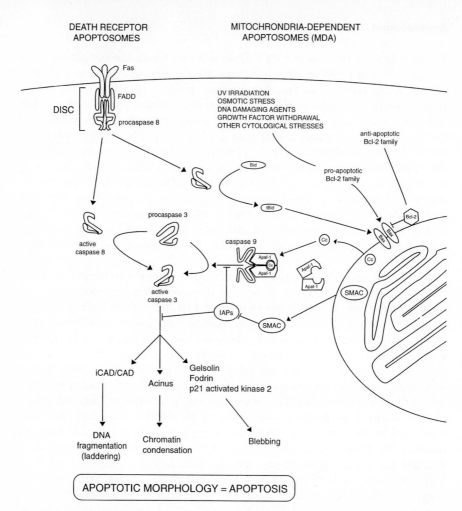

Fig. 2. Schematic representation of major steps in the formation of two kinds of apoptosomes. On the left, the death receptor apoptosome forms a death-inducing signaling complex (DISC) around a ligand-bound death receptor, in this case Fas. Transactivation of procaspase-8 leads to its cleavage of procaspase-3 and the pro-apoptotic Bcl-2 family member Bid. Truncated Bid (tBID) translocates to the mitochondria where it facilitates in the release of cytochrome *c*. On the right, pro-apoptotic stimuli can trigger the release of cytochrome *c* through the actions of Bcl-2 family members. Cytoplasmic cytochrome *c* can interact with Apaf-1 to form a complex that can then recruit and activate procaspase-9. Active caspase-9, within this complex, quickly cleaves procaspase-3. Inhibition of caspases by IAPs is eliminated by the release of SMAC from mitochondria, which occurs synchronously with cytochrome *c* release. The activation of caspase-3 cleaves substrates that effect the morphologic changes associated with apoptosis

believed that cytochrome *c* binds to the WD repeats of Apaf-1, causing a conformational change that allows it to oligomerize and interact with caspase-9. As the only major non-cytoplasmic component of MDA, cytochrome *c* release from mitochondria is an important regulatory step in this pathway.

Within the nervous system we have found that MDA are responsible for the apoptotic component of NMDA-mediated excitotoxicity (Ethell et al., in preparation). Moderate stimulation of NMDA-sensitive glutamate receptors in vitro leads to cytochrome c that is caspase independent (i.e., upstream). In addition, the subsequent apoptosis can be greatly blocked by the pan-caspase inhibitor zVAD-fmk. Furthermore, we have found that cerebrocortical neurons cultured from Apaf-1 deficient mouse embryos are resistant to NMDA-mediated apoptosis. Our findings suggest that the development of inhibitors to the formation and activation of MDA may alleviate apoptosis resulting from NMDA-receptor overstimulation, without altering physiological glutamate signaling.

Factors Affecting Cytochrome c Release

Mature holo-cytochrome c is covalently linked with heme and forms a globular structure that cannot go through the intact outer mitochondrial membrane (OMM). It lies trapped in the intermembrane space of mitochondria, where it shuttles electrons between complexes III and IV of the electron transport chain. The million dollar question is: how does cytochrome c get out to the cytoplasm where it can interact with Apaf-1? Before we get to some current models of this process, we will first review the Bcl-2 family of proteins, as they play a fundamental role in the regulation of this process. Similarities between Bcl-2 family members are highest in four different alpha helical segments called Bcl-2 homology domains (BH1-4). Anti-apoptotic Bcl-2 family members generally contain all four BH domains and include CED-9, Bcl-2, Bcl-XL, Bcl-W, MCL-1 and A1/BFL-1 (Gross et al. 1999). NR-13 and BOO/DIVA are anti-apoptotic members that lack the BH3 domain. Indeed BH3 has been implicated as critical to the pro-apoptotic activity of the Bcl-2 family members Bax, Bak and Bok/MTD, which all contain BH1-3. The recent discovery of several BH3-only members – Bid, Bad, Bik, BLK, BIM/NBK, BIM/BOD, HRK, NIP3, BNIP3 and EGL-1 – has strengthened this argument.

It has been suggested that Bcl-2, Bcl-XL and other anti-apoptotic Bcl-2 family members can block the function of pro-apoptotic family members by heterodimerization. Conversely, the anti-apoptotic members can be overpowered by greater numbers of pro-apoptotic proteins. This reasoning has led to the concept of molar balances between the pro- and anti-apoptotic members, which will alter the susceptibility of cells to apoptotic stimuli, depending on which way the scale tips. For the sake of arguement and brevity let us say that Bax has overcome Bcl-2 and is now free to release cytochrome c. But how does it do that?

As with all Bcl-2 family proteins, Bax has channel-forming properties when inserted into a membrane. One popular model is that many Bax proteins can come together to form large enough pores in the OMM for cytochrome c to escape, although this idea remains unproven. Alternatively, it has been suggested that Bax can hi-jack large channel proteins that reside in the OMM and change them to allow cytochrome c to escape. One such protein, TOM40, is part of a protein junction between the inner mitochondrial membrane (IMM) and the OMM. Although the channel pore of TOM40 is big enough to allow cytochrome c free passage, this theory has not been proven one way or another. Another protein that Bax may change in the OMM to allow cytochrome c release is the voltage-dependent anion channel (VDAC), which usually complexes with the adenine nucleotide transporter (ANT) on the IMM.

Another model suggests that cytochrome c release is a consequence of disruption of the ionic channels across the IMM, resulting in what is called a permeability transition (PT). In the PT model, mitochondria begin to take up water into the inner mitochondrial matrix, which swells until it reaches the rigid OMM: Eventually the pressure becomes so great that it ruptures the OMM and squeezes out cytochrome c. However, the PT model predicts a loss of potential across the IMM in response to pro-apoptotic Bcl-2 family members. Newmeyer and colleagues have found this not to be the case in several situations. First, isolated mitochondria that have released cytochrome c in response to tBID are still capable of importing proteins into the inner mitochondrial space with the TIM/TOM complex, and this can only happen in the presence of a substantial membrane potential ($\Delta\psi$) across the IMM (von Ahsen 2000). Further, by looking at cells that express GFP-cytochrome c, we have seen those cells lose cytochrome c but retain their mitochondrial membrane potential, as indicated by the dye TMRE (Goldstein et al. 2000). Importantly, we should note that this maintenance of membrane potential only occurs in the presence of the caspase inhibitor zVAD-fmk. We have previously shown cytochrome c release to be caspase independent (Bossy-Wetzel 1998), but this new finding suggests that once caspases are active they return to the mitochondria and cause a loss of $\Delta\psi$. Mechanisms of cytochrome c release remain one of the most hotly debated areas in apoptosis research; however, as we begin to understand the interplay of Bcl-2 family members, we will eventually clear up this critical step in apoptosis.

IAPs and SMAC/DIABLO

In addition to caspases, cells also contain natural inhibitors of the caspases. These inhibitor of apoptosis proteins (IAPs) were first identified in baculovirus (p35) but were subsequently found in human cells (Deveraux and Reed 1999). The IAPs all contain domains homologous to the baculoviral IAP repeat (BIR domains), which can occur once, as with survivin, or in three tandem repeats, as with XIAP, NAIP and cIAPS 1 and 2. However, in XIAP it has been shown that only the middle BIR2 domain is sufficient to inhibit caspases 3, 7 and 9 (Deveraux et al. 1997, 1998). Another domain found in XIAP and cIAPs 1 and 2 is a carboxyl-Zn^{2+} ring finger that is thought to aid in the proteosomal degradation of inhibited caspases. However, the ring domains are not necessary to block caspase activity; that action seems to lie exclusively with the BIR domains. Interestingly, cIAPs also contain caspase recruitment domains (CARD) between the last tandem BIR repeat and the Zn^{2+} finger. Though the function of these CARDs has not been established, it is intriguing to speculate how they share the same protein-protein interaction motif as caspase-9 and Apaf-1. IAPs are the first detected family of endogenous mammalian proteins that inhibit caspases, yet much remains to be learned about their regulation.

A recent advancement with regard to IAP regulation is the cloning and characterization of an IAP inhibitor SMAC/DIABLO (Du et al. 2000; Verhagen et al. 2000). This protein is translated in an inactive proform that is translocated into mitochondria where the amino terminal 55 amino acids are cleaved off to leave active SMAC within. When apoptotic stimuli trigger cytochrome c release, SMAC is also released into the cytoplasm. Once released from mitochondria, SMAC is free to interact with IAPs and inhibit them. In essence SMAC disinhibits the caspases. The discovery of SMAC and its function bridges the gap between apoptosis in mammalian and insect cells. For years we have been puzzled by apoptosis in insect cells as they do not appear to regulate apoptosis via the release of cytochrome c, but more simply through the positive and negative regulation of caspase activities with molecules like RPR, HID and GRIM that disinhibit IAPs, just like SMAC.

But When You Tie It All Together, You Get Into Knots

To demonstrate how apoptosis research can provide insights into neurological research, and also how far we have yet to go, let us briefly consider apoptosis initiated by neurotrophin withdrawal. In addition to NGF, other neurotrophins – BDNF NT-3, NT-4/5 and GDNF – will support the survival of different neuronal populations. NGF binding to its high affinity receptor, trkA, results in receptor dimerization and autophosphorylation of the cytoplasmic tyrosine kinase domain. Among the signals generated in response to the activation of these kinases is increased activity of PhosphatidylInositol-3-kinase (PI-3-K; Kapeller and Cantley 1994). Membrane localization of the PI-3-K regulatory subunit (p85) leads to co-localization and subsequent activation of the p110 catalytic subunit. PI-3-K phosphorylates the lipid phosphatidylInositol-3-P (PtdIns-3-P) to form PtdIns-3,4-P2 and PtdIns-3,4,5-P3, which also forms PtdIns-3,4-P2 by the removal of the phosphate in position 5. Among the many pathways affected by different levels of these PtdIns forms is the serine/threonine kinase Akt (PKB). Binding of PtdIns-3,4-P2 by the Akt pleckstrin homology (PH) domain facilitates juxtamembrane localization, homodimerization and phosphorylation at T308 and S427 by PRK1/2 (Franke et al. 1997; Klippel et al. 1997; Delcommenne et al. 1998). Activation of this PI-3-K/Akt pathway has been shown to prevent apoptosis induced by UV irradiation in COS-7 and Rat-1 cells (Kulik et al. 1997), anoikis in MDCK cells (Khwaja et al. 1997), deregulated c-myc in Rat-1 cells (Kauffman-Zeh et al. 1997; Kennedy et al. 1997) and survival factor (neurotrophin) withdrawal in neurons (D'Mello et al. 1997; Dudek et al. 1997).

The ways in which Akt activity may save neurons from apoptotis is by preventing formation of either MDA, death receptors apoptosomes, or both. One possible mechanism is through Akt phosphorylation of the pro-apoptotic Bcl-2 family member Bad (Datta et al. 1997; del Peso et al. 1997), which has then sequestered by the cytoplasmic protein 14-3-3 (Zha et al. 1996). Dephosphorylation of Bad by calcineurin allows it to dissociate from 14-3-3 and translocate to mitochondria (Wang et al. 1999). However, at the mitochondrial outer membrane Bad may be re-phosphorylated by mitochondria-anchored protein kinase A, which would prevent its actions (Harada et al. 1999). Along these same lines, Deckwerth et al. (1996) reported that sympathetic neurons derived from mice deficient for the pro-apoptotic Bcl-2 family member, Bax, are resistant to NGF withdrawal. Together these reports suggest that neurotrophins support neuronal survival by inhibiting cytochrome *c* release and hence MDA activation, yet the relative importance of specific proteins remains controversial.

However, complicating this explanation is a recent report that Akt activity supports cerebellar neurons from BDNF withdrawal by phosphorylating, and inactivating, a forkhead transcription factor (Brunet et al. 1999). These researchers suggest that decreased Akt activity leads to net dephosphorylation of this forkhead transcription factor, which becomes active and directs the expression of FasL. As discussed previously, expression of FasL can lead to cis and trans activation of Fas, which will activate caspase-8 through a death receptor apoptosome.

Concluding Remarks

The investigation of apoptotic mechanisms is currently one of the most active areas for basic and applied research. This newfound knowledge is of particular relevance to the neurodegenerative disorders since we can, hopefully, trace backwards from how the neurons die to why. Currently, the regulation of cytochrome *c* release by members of the Bcl-2 family is a major bottleneck to going further back. However, within the next few years we will almost certainly move beyond that stage and learn the upstream regulatory events require particular attention.

References

Alnemri ES, Livingston DJ, Nicholson DW, Salvesen G, Thornberry NA, Wong WW, Yuan J (1996) Human ICE/CED-3 protease nomenclature. Cell 87(2):171

Bossy-Wetzel E, Newmeyer DD, Green DR (1998) Mitochondrial cytochrome *c* release in apoptosis occurs upstream of DEVD-specific caspase activation and independently of mitochondrial transmembrane depolarization. EMBO J 17:37–49

Brunet A, Bonni A, Zigmond MJ, Lin MZ, Juo P, Hu LS, Anderson MJ, Arden KC, Blenis J, Greenberg ME (1999) Akt promotes cell survival by phosphorylating and inhibiting a Forkhead transcription factor. Cell 96(6):857–868

Collins RJ, Harmon BV, Gobe GC, Kerr JF (1992) Internucleosomal DNA cleavage should not be the sole criterion for identifying apoptosis. Int J Radiat Biol 61:451–453

Cotter TG, Lennon SV, Glynn JM, Green DR (1992) Microfilament-disrupting agents prevent the formation of apoptotic bodies in tumor cells undergoing apoptosis. Cancer Res 52:997–1005

D'Mello SR, Borodezt K, Soltoff S (1997) Insulin-like growth factor and potassium maintain neuronal survival by distinct pathways: possible involvement of PI 3-kinase in IGF-1 signaling. J Neurosci 17(5):1548–1560

Datta SR, Dudek H, Tao X, Masters S, Fu H, Gotoh Y, Greenberg ME (1997) Akt phosphorylation of BAD couples survival signals to the cell intrinsic death machinery. Cell 91(2):231–241

Deckwerth TL, Elliott JL, Knudson CM, Johnson EM Jr, Snider WD, Korsmeyer SJ (1996) BAX is required for neuronal death after trophic factor deprivation and during development. Neuron 17(3):401–411

Delcommenne M, Tan C, Gray V, Rue L, Woodgett J, Dedhar S (1998) Phosphoinositide-3-OH kinase-dependent regulation of glycogen synthase kinase 3 and protein kinase B/Akt by the integrin-linked kinase. Proc Natl Acad Sci USA 95:11211–11216

del Peso L, Gonzalez-Garcia M, Page C, Herrerra R, Nunez G (1997) Interleukin-3-induced phosphorylation of BAD through the protein kinase Akt. Science 278(5338):687–689

Deveraux QL, Reed JC (1999) IAP family proteins – suppressors of apoptosis. Genes Dev 13:239–252

Deveraux QL, Roy N, Stennicke HR, Van Arsdale T, Zhou Q, Srinivasula SM, Alnemri ES, Salvesen GS, Reed JC (1998) IAPs block apoptotic events induced by caspase-8 and cytochrome c by direct inhibition of distinct caspases. EMBO J 17(8):2215–2223

Deveraux QL, Takahashi R, Salvesen GS, Reed JC (1997) X-linked IAP is a direct inhibitor of cell-death proteases. Nature 388(6639):300–304

Devitt A, Moffatt OD, Raykundalia C, Capra JD, Simmons DL, Gregory CD (1998) Human CD14 mediates recognition and phagocytosis of apoptotic cells. Nature 392:505–509

Du C, Fang M, Li Y, Li L, Wang X (2000) Smac, a mitochondrial protein that promotes cytochrome c-dependent caspase activation by eliminating IAP inhibition. Cell 102(1):33–42

Dudek H, Datta SR, Franke TF, Birnbaum MJ, Yao R, Cooper GM, Segal RA, Kaplan DR, Greenberg ME (1997) Regulation of neuronal survival by the serine-threonine protein kinase Akt. Science 275:661–665

Enari M, Sakahira H, Yokoyama H, Okawa K, Iwamatsu A, Nagata S (1998) A caspase-activated DNase that degrades DNA during apoptosis, and its inhibitor ICAD. Nature 391:43–50

Fadok VA, Warner ML, Bratton DL, Henson PM (1998) CD36 is required for phagocytosis of apoptotic cells by human macrophages that use either a phosphatidylserine receptor or the vitronectin receptor (alpha v beta 3). J Immunol 161:6250–6257

Fadok VA, Bratton DL, Rose DM, Pearson A, Ezekewitz RAB, Henson PM (2000) A new receptor for phosphatidylserine-specific clearance of apoptotic cells. Nature, in press

Franke TF, Kaplan CR, Cantley LC, Toker A (1997) Direct regulation of the Akt proto-oncogene by phosphatidylinositol-3,4-biphosphate. Science 275:665–668

Goldstein JC, Waterhouse NJ, Juin P, Evan GI, Green DR (2000) The coordinate release of cytochrome a c during apoptosis is rapid, complete and kinetically invariant. Nat Cell Biol 2(3):156–162

Green DR (1998) Apoptotic pathways: the roads to ruin. Cell 94:695–698

Gross A, McDonnell JM, Korsmeyer SJ (1999) BCL-2 family members and the mitochondria in apoptosis. Genes Dev 13(15):1899–1911

Harada H, Becknell B, Wilm M, Mann M, Huang LJ, Taylor SS, Scott JD, Korsmeyer SJ (1999) Phosphorylation and inactivation of BAD by mitochondrial-anchored protein kinase A. Mol Cell 3:413–422

Kapeller R, Cantley LC (1994) Phosphatidylinositol 3-kinase. Bioessays 16:565–576

Kauffman-Zeh A, Rodriguez-Viciana P, Ulrich E, Gilbert C, Coffer P, Downward J, Evan G (1997) Supression of c-myc-induced apoptosis by Ras signalling through PI(3)K and PKB. Nature 385(6616):544–548

Kennedy SG, Wagner AJ, Conzen SD, Jordan J, Bellacosa A, Tsichlis PN (1997) The PI 3-kinase/Akt signaling pathway delivers an anti-apoptotic signal. Genes Dev 11(6):701–713

Khwaya A, Rodriguez-Viciana P, Wennstrom S, Warne PH, Downward J (1997) Matrix adhesion and Ras transformation both activate a phosphoinositide 3-OH kinase and protein kinase B/Akt cellular survival pathway. EMBO J 16(10):2783–2793

Klippel A, Kavanaugh WM, Pot D, Williams LT (1997) A specific product of phosphatidylinositol 3-kinase directly activates the protein kinase Akt through its pleckstrin homology domain. Mol Cell Biol 17:338–344

Kothakota S, Azuma T, Reinhard C, Klippel A, Tang J, Chu K, McGarry TJ, Kirschner MW, Koths K, Kwiatkowski DJ, William LT (1977) Caspase-3-generated fragment of gelsolin: effector of morphological change in apoptosis. Science 278:294–298

Kulik G, Klippel A, Weber MJ (1997) Antiapoptotic signalling by the insulin-like growth factor receptor, phosphatidylinositol 3-kinase, and Akt. Mol Cell Biol 17(3):1595–1606

Lee N, McDonald H, Reinhard C, Halenbeck R, Roulston A, Shi T, Williams LT (1997) Activation of hPAK65 by caspase cleavage indues some of the morphological and biochemical changes of apoptosis. Proc Natl Acad Sci USA 94:13642–13647

Le-Niculescu H, Bonfoco E, Kasuya Y, Claret FX, Green DR, Karin M (1998) Withdrawal of survival factors results in activation of the JNK pathway in neuronal cells leading to Fas ligand induction and cell death. Mol Cell Biol 19(1):751–763

Liu X, Zou H, Slaugther C, Wang X (1997) DFF, a heterodimeric protein that functions downstream of caspase-3 to trigger DNA fragmentation during apoptosis. Cell 89:175–184

Martin SJ, O'Brien GA, Nishioka WK, McGahon AJ, Mahboubi A, Saido TC, Green DR (1995) Proteolysis of fodrin (non-erythroid spectrin) during apoptosis. J Biol Chem 270:6425–6428

Martin SJ, Reutelingsperger CP, McGahon AJ, Rader JA, van Schie RC, LaFace DM, Green DR (1995) Early redistribution of plasma membrane phosphatidylserine is a general feature of apoptosis regardless of the initiating stimulus: inhibition by overexpression of Bcl-2 and Abl. J Exp Med 182:1545–1556

Martin SJ, Finucane DM, Amarante-Mendes GP, O'Brien GA, Green DR (1996) Phosphatidylserine externalization during CD95-induced apoptosis of cells and cytoplasts requires ICE/CED-3 protease activity. J Biol Chem 271:28753–28756

McCarthy NJ, Whyte MK, Gilbert CS, Evan GI (1997) Inhibition of Ced-3/ICE-related proteases does not prevent cell death induced by oncogenes, DNA damage, or the Bcl-2 homologue Bak. J Cell Biol 136:215–227

Ren Y, Silverstein RL, Allen J, Savill J (1995) CD36 gene transfer confers capacity for phagocytosis of cells undergoing apoptosis. J Exp Med 181:1857–1862

Rudel T, Bokoch GM (1997) Membrane and morphological changes in apoptotic cells regulated by caspase-mediated activation of PAK2. Science 276:1571–1574

Sahara S, Aoto M, Eguchi Y, Imamoto N, Yoneda Y, Tsujimoto Y (1999) Acinus is a caspase-3-activated protein required for apoptotic chromatin condensation. Nature 401:168–173

Savill J, Dransfield I, Hogg N, Haslett C (1990) Vitronectin receptor-mediated phagocytosis of cells undergoing apoptosis. Nature 343:170–173

Stennicke HR, Salvesen GS (1999) Catalytic properties of the caspases. Cell Death Diff 6(11):1054–1059

Thornberry NA, Lazebnik Y (1998) Caspases: enemies within. Science 281:1312–1316

Vanags DM, Porn-Ares MI, Coppola S, Burgess DH, Orrenius S (1996) Protease involvement in fodrin cleavage and phosphatidylserine exposure in apoptosis. J Biol Chem 271:31075–31085

Verhagen AM, Ekert PG, Pakusch M, Silke J, Connolly LM, Reid GE, Moritz RL, Simpson RJ, Vaux DL (2000) Identification of DIABLO, a mammalian protein that promotes apoptosis by binding to and antagonizing IAP proteins. Cell 102:43–53

von Ahsen O, Renken C, Perkins G, Kluck RM, Bossy-Wetzel E, Newmeyer DD (2000) Preservation of mitochondrial structure and function after Bid- or Bax-mediated cytochrome c release. J Cell Biol 150:1027–1036

Wang HG, Pathan N, Ethell IM, Krajewski S, Yamaguchi Y, Shibasaki F, McKeon F, Bobo T, Franke TF, Reed JC (1999) Ca^{2+}-induced apoptosis through calcineurin dephosphorylation of BAD. Science 284(5412):339–343

Weil M, Jacobson MD, Coles HS, Davies TJ, Gardner RL, Raff KD, Raff MC (1996) Constitutive expression of the machinery for programmed cell death. J Cell Biol 133(5):1053–1059

Wolf BB, Green DR (1999) Suicidal tendencies: apoptotic cell death by caspase family proteinases. J Biol Chem 274:20049–20052

Wolf BB, Goldstein JC, Stennicke HR, Beere H, Amarante-Mendes GP, Salvesen GS, Green DR (1999) Calpain functions in a caspase-independent manner to promote apoptosis-like events during platelet activation. Blood 94:1683–1692

Zha J, Harada H, Yang E, Jockel J, Korsmeyer SJ (1996) Serine phosphorylation of death agonist BAD in response to survival factor results in binding to 14-3-3 not BCL-X(L). Cell 87:619–628

Zhang J, Lee H, Lou DW, Bovin GP, Xu M (2000) Lack of obvious 50 kilobase pair DNA fragments in DNA fragmentation factor 45-deficient thymocytes upon activation of apoptosis. Biochem Biophys Res Commun 274(1):225–229

Zou H, Henzel WJ, Liu X, Lutschg A, Wang X (1997) Apaf-1, a human protein homologous to C. elegans CED-4, participates in cytochrome c-dependent activation of caspase-3. Cell 90(3):405–413

Mitochondrial Membrane Permeabilization in Physiological and Pathological Cell Death

L. Ravagnan and G. Kroemer

Summary

Mitochondrial membrane permeabilization (MMP) is a hallmark of early apoptosis, including apoptosis occurring in neurons. Evidence is accumulating that mitochondria play a prime role in the processes of ischemia/reperfusion damage, excitotoxicity and neurodegeneration. MMP is regulated by pro- and anti-apoptotic members of a Bax/Bcl-2 family of proteins, via a process that may involve sessile mitochondrial proteins organized in the two membrane-spanning permeability transition pore complex (PTPC). Apoptotic MMP differentially affects the outer and inner mitochondrial membranes. The inner mitochondrial membrane becomes permeable to solutes up to 1500 Da, yet retains matrix proteins in their normal localization. In contrast, the outer mitochondrial membrane becomes completely permeabilized to proteins, leading to the release of soluble proteins from the mitochondrial intermembrane space to an ectopic (extramitochondrial) localization. Several intermembrane proteins can activate catabolic hydrolases involved in the apoptotic process. One such protein is cytochrome c, which participates in the caspase activation cascade. Another functionally important intermembrane protein is apoptosis inducing factor (AIF), which provokes nuclear DNA fragmentation and chromatin condensation via a caspase-independent process. In addition to its local nuclear effects, AIF may also participate in the overall regulation of apoptosis, since neutralization of AIF by microinjection of a specific antibody can abort MMP and subsequent apoptosis, at least in some models. The relevance of AIF for caspase-independent neuronal cell death remains to be established.

Mitochondrial Membrane Permeabilization: An Early Feature of Cell Death

Mitochondrial membrane permeabilization (MMP) is a near-to-general phenomenon associated with apoptosis (Green and Reed 1998; Green and Kroemer 1998; Kroemer and Reed 2000). With respect to MMP, three phases of the apoptotic process can be distinguished. During a pre-mitochondrial phase, pro-apoptotic second messengers are activated and factors acting on mitochondrial membranes accumulate and/or translocate to mitochondria. Prominent pro-apoptotic proteins directly acting on mitochondria include

Henderson/Green/Mariani/Christen (Eds.)
Neuronal Death by Accident or by Design
© Springer-Verlag Berlin Heidelberg 2001

the pro-apoptotic members of the Bcl-2/Bax family, as well as kinases and phosphatases that induce covalent modification of Bcl-2/Bax family members, thereby influencing their function and/or subcellular localization. The transcription factors Nur77/NGFI-B and p53 have also recently been shown to translocate to mitochondrial membranes and to permeabilize them (Li et al. 2000; Marchenko et al. 2000). Non-protein factors acting on mitochondria include Ca^{2+}, reactive oxygen species, NO, and ganglioside GD3, as well as some experimental anti-cancer agents (Kroemer 1997; Gross et al. 1999; Vander Heiden and Thompson 1999; Kroemer and Reed 2000; Ravagnan et al. 1999). During the mitochondrial phase, MMP occurs, presumably through a limited set of mechanisms. Finally, during the post-mitochondrial phase, the functional consequences of MMP, namely bioenergetic failure and/or release of potentially harmful proteins normally sequestered in mitochondria, culminate in the cell death.

The signs of MMP include the release of soluble proteins from the mitochondrial intermembrane space [e.g., cytochrome c, apoptosis inducing factor (AIF), Smac/Diablo] through the outer membrane (OM), as well as a partial permeabilization of the inner membrane (IM) for solutes up to ~ 1500 Da, leading to a reduction of the mitochondrial transmembrane potential ($\Delta\Psi_m$). OM permeabilization can be monitored by confocal scanning microscopy, either after immunofluorescence staining (Green and Reed 1998; Susin et al. 1999) or after transfection of cells with chimerica cDNAs coding for intermembrane proteins (e.g., cytochrome c or AIF) fused to green fluorescent protein (GFP; Heiskanen et al. 1999; Goldstein et al. 2000; Loeffler et al. 2001). In some models of apoptosis, electron microscopy reveals local ruptures of the OM with herniation of the IM (Vander Heiden and Thompson 1999; Kwong et al. 1999). IM permeabilization can be measured in intact cells by loading the cytosol or the mitochondrial matrix with calcein, a hydrophilic 620 Da fluorochrome that normally does not cross the IM (Lemasters et al. 1998; Bernardi et al. 1999). Alternatively, IM permeabilization can be assessed indirectly by determining a reduction in the $\Delta\Psi_m$. To this end, cells are incubated with lipophilic cationic fluorochromes such as $DiOC_{6(3)}$ or JC-1, which accumulate in the mitochondrial matrix, driven by the $\Delta\Psi_m$. A reduction in fluorescence intensity (the case of $DiOC_{6(3)}$) or a concentration-dependent shift in the emission spectrum (red → green in the case of JC-1) is then intepreted as a sign of $\Delta\Psi_m$ dissipation (Kroemer et al. 1998; Bernardi et al. 1999). IM permeabilization is a less constant feature of apoptosis than OM permeabilization. Although cytochrome c release through OM is mostly associated with a permanent loss of $\Delta\Psi_m$ (Heiskanen et al. 1999), this $\Delta\Psi_m$ reduction may be transient, indicating IM "resealing" (Petronilli et al. 1999; Pastorino et al. 1999; Szalai et al. 1999). Moreover, in some cases cytochrome c release has been reported to occur in cells whose mitochondria have an apparently normal (Green and Reed 1998; Goldstein et al. 2000) or even an increased $\Delta\Psi_m$ (Van der Heiden et al. 1999).

Mechanisms of MMP

It appears that MMP is induced by at least two major mechanisms. One involves the mitochondrial permeability transition pore complex (PTPC), also called mitochondrial megachannel, whereas the other appears to be PTPC-independent. The PTPC is a multiprotein complex formed at the contact site between the mitochondrial IM and OM. The core components of the PTPC are the adenine nucleotide translocase (ANT, in the inner membrane), cyclophilin D (in the matrix) and the voltage-dependent anion channel (VDAC, in the OM (Marzo et al. 1998b; Crompton et al. 1998). Five independent lines of evidence implicate PTPC in the regulation of cell death.

- First, in intact cells, apoptosis is accompanied by an early permeabilization of mitochondrial membranes (Zamzami et al. 1995a,b). Several PTPC-inhibitory agents, including ligands of ANT (bongkrekate), cyclophilin D (cyclosporin A and its non-immunosuppressive derivative N-methyl-4-Val-cyclosporin) and VDAC (Koenig's polyanion), prevent mitochondrial membrane permeabilization (MMP) and subsequent cell death (Marchetti et al. 1996; Ferri et al. 2000b; Kroemer and Reed 2000).
- Second, mitochondria are rate-limiting for caspase and nuclease activation in cell-free systems of apoptosis. Isolated mitochondria release apoptogenic factors capable of activating pro-caspases or endonucleases upon opening of the PTPC in vitro (Zamzami et al. 1996; Liu et al. 1996; Budijardjo et al. 1999; Susin et al. 1999).
- Third, opening of the purified PTPC reconstituted into liposomes in inhibited by recombinant Bcl-2 or Bcl-XL, two apoptosis-inhibitory proteins that also prevent PTPC opening in cells and isolated mitochondria (Marzo et al. 1998b). Bcl-2 also inhibits pore formation by ANT (Marzo et al. 1998a; Brenner et al. 2000) and VDAC (Shimizu et al. 1999, 2000), both in proteoliposomes and in synthetic lipid bilayers subjected to electrophysiological measurements.
- Fourth, pro-apoptotic proteins such as Bax may act on PTPC consituents, including ANT and VDAC, to facilitate MMP. Again, this has been shown in biochemically defined systems including liposomes and plain lipid bilayers containing Bax, ANT, and/or VDAC (Marzo et al. 1998a; Brenner et al. 2000; Shimizu et al. 1999, 2000).
- Fifth, a number of endogenous, viral, or xenogeneic effectors directly act on the PTPC to trigger the apoptotic cascade (Jacotot et al. 2000; Ferri et al. 2000a; Kroemer and Reed 2000; Ravagnan et al. 1999; Costantini et al. 2000).

Altogether, these findings suggest that PTPC opening may be sufficient and necessary for triggering apoptosis in a variety of different models. However, a number of authors have recently suggested the existence of PTPC-independent modes of mitochondrial membrane permeabilization. One particularly well-documented example concerns the action of Bax-like members of the Bcl-2 family (Saito et al. 2000). Aided by the interaction with truncated Bid (t-Bid), Bax and Bak may insert into the OM while undergoing oligomerization, thereby forming a protein-permeable conduit (Eskes et al. 2000; Wei et al. 2000). Reportedly, this type of OM permeabilization does not require interaction with proteins from the PTPC and is not affected by pharmacological PTPC inhibitors (Desagher et al. 1999; Eskes et al. 2000; Wei et al. 2000). At present it is not clear under which specific circumstances the PTPC-dependent and PTPC-independent mechanism of MMP prevail. This may depend on the apoptosis-inducing stimulus. In addition, cell type-specific differences have been reported. Thus, the mechanisms of MMP induced in vitro appear to be predominantly PTPC-independent in isolated brain mitochondria [which must not reflect the situation in intact cells (Petersen et al. 2000)], whereas liver mitochondria preferentially permeabilize in a PTPC-dependent fashion (Andreyev and Fiskum 1999; Berman et al. 2000).

Mitochondrial Involvement in Acute and Chronic Neuronal Death

Transient brain ischemia is accompanied by acute or delayed cell death, which may involve regulatory events acting the level of MMP. Studies of cultured cells revealed that transient deprivation of oxygen and/or nutrients can trigger MMP and cell death, while cyclosporin A (CsA) can prevent MMP and consequent cell death (Lemasters et al. 1998). CsA blocks calcineurin (which indirectly affects the pro-apoptotic Bcl-2 family member Bad; Wang et al. 1999) as well as the cyclophilin D/ANT interaction (Crompton 1999) and thus may employ two mechanisms to reduce MMP. Either FK506 (which acts on calcineurin but not on cyclophilin D) or the nonimmunosuppressive CsA analog N-Methyl-Val-4-CsA (which only acts on cyclophilin D) may confer neuroprotection, depending on the model that is studied (Khaspekov et al. 1999). In vivo experiments reveal that CsA can prevent MMP and reduce infarction size in brain ischemia (Yoshimoto and Siesjo 1999). CsA also prevents cell death-related mitochondrial alterations in hypoglycemia-induced hippocampal damage (Ferrand Drake et al. 1999) or trauma-induced cortical damage (Okonkwo et al. 1999). In vitro, CsA can inhibit the MMP and cell death induced by the excitotoxin glutamate. Moreover, CsA antagonizes MMP induced by the parkinsonian neurotoxin N-methyl-4-phenylpyridinium (MPP^+; Cassarino et al. 1999) or by the dopamine oxidation product dopamine quinone (Berman and Hastings 1999) studied on isolated brain mitochondria.

Several neurodegenerative diseases caused by defects in nuclear genes affecting mitochondrial functions are clearly linked to the pathological cell death of post-mitotic neurons (Table 1). The mitochondrial etiology has been established for Friedreich's ataxia (FRDA) and hereditary spastic paraplegia, in which two nuclear-encoded mitochondrial proteins, frataxin and paraplegin, respectively, are mutated (Koutnikova et al. 1997; Casari et al. 1998). How these mutations relate to cell death remains elusive. Huntingtin is the protein that is mutated by extension of a polyglutamine tract in most

Table 1. Evidence for a link between neurodegeneration and mitochondrial dysfunction

Disease	Observation	References
Friedreich's ataxia (FRDA)	Frataxin, the protein mutated in FRDA, localizes to mitochondria and is required to maintain mitochondrial iron homestasis and DNA content.	Koutnikova et al. 1997
Hereditary spastic paraplegia	Mutations in paraplegin, a mitochondrial metalloprotease, cause defects on oxidative phosphorylation.	Casari et al. 1998
Huntington's disease (HD)	Treatment with 3-nitropropionic acid, a respiratory chain complex II inhibitor, causes an HD-like disease in primates and rodents.	Brouillet et al. 1999 Leventhal et al. 2000
	Peripheral lymphoblasts from patients with HD exhibit enhanced cyanide-induced $\Delta\Psi_m$ loss and apoptosis susceptibility correlated with increased glutamine repeats. $\Delta\Psi_m$ reduction is corrected by CsA.	Sawa et al. 1999
Amyotrophic lateral sclerosis (ALS)	Overexpression of human mutant SOD-1 G93A causes $\Delta\Psi_m$ dissipation in neuroblastoma cells. Bcl-2, coenzyme Q, and creatine attenuate a transgenic mouse model of ALS.	Carri et al. 1997 Kostic et al. 1997 Matthews et al. 1998 Klivenyi et al. 1999
Parkinson's disease (PD)	Intoxication with 1-methyl-4-phenylpyridinium, a complex I inhibitor and MMP inducer, causes PD in humans.	Cooper and Schapira 1997
	Cybrids generated with mitochondra from PD patients have reduced complex I activity, reduced protonophore-releasable calcium, and enhanced susceptibility to apoptosis induction by 1-methyl-4-phenylpyridium.	Sheehan et al. 1997
Alzheimer's disease (AD)	Mitochondrial DNA deletions and mutations in Alzheimer's brains.	de la Monte et al. 2000
	Cybrids generated with mitochondria from sporadic AD patients exhibit a reduced $\Delta\Psi_m$ that is corrected by CsA.	Cassarino et al. 1998
	Mutations in presenilin-1 increase the vulnerability of neural cells to mitochondrial toxins (nitropropironic acid or malonate) and increase amyloid β-induced mitochondrial dysfunction.	Guo et al. 1999
	Presenilin-1 and -2 interact with Bcl-2.	Passer et al. 1999

cases of Huntington's disease (HD). Lymphoblasts from HD patients are abnormally susceptible to $\Delta\Psi_m$ loss and cell death induced by apoptogenic stimuli such as staurosporin and two respiratory chain complex IV inhibitors (cyanide and azide). The extent of cyanide-induced $\Delta\Psi_m$ loss correlates with the length of the polyglutamine tract of huntingtin. Moreover, $\Delta\Psi_m$ dissipation is prevented by CsA (but not by the calcineurin inhibitor FK506; Sawa et al. 1999), implicating the PTPC in the abnormal apoptosis susceptibility of

HD lymphocytes. Circumstantial evidence for mitochondrial involvement is also available for Alzheimer's disease (Table 1) and amyotrophic lateral sclerosis (ALS). Transgenic overexpression of Bcl-2, as well as treatment with the MMP inhibitors coenzyme Q (ubiquinone; Fontaine et al. 1998) or creatine, can prolong the life span of mice carrying a SOD-1 mutation found in patients with familial ALS (Kostic et al. 1997; Matthews et al. 1998; Klivenyi et al. 1999). Creatine also has a beneficial effect on a mouse model of HD (Ferrante et al. 2000). These latter observations underscore the possibility that neuroprotection may be achieved by pharmacologically intervening on mitochondria.

AIF as a Novel Mitochondrial Death Effector

As a result of MMP, proteins normally confined to the intermembrane space are translocated to non-mitochondrial structures (Liu et al. 1996; Susin et al. 1996; Patterson et al. 2000). This translocation then triggers catabolic reactions that give rise to the apoptotic phenotype. As an example, cytochrome *c* redistributes from the mitochondrion to the cytosol (Liu et al. 1996) and triggers the formation of a caspase-9/caspase-3 activating complex, the apoptosome. We have recently identified a caspase-independent death effector, AIF, that is also an intermembrane space protein. AIF is strongly conserved among mammalian species (92% aa identity in the whole protein between mouse and man). The mouse AIF cDNA codes for a protein that is organized in three domains: 1) an amino-terminal mitochondrial localization sequence (MLS) of 100 amino acids; 2) a spacer sequence of 27 aminoacids; and 3) a carboxyterminal 485 amino acid domain with strong homology to oxidoreductases from other vertebrates (*X. laevis*), non-vertebrate animals (*C. elegans, D. melanogaster*), plants, fungi, eubacteria, and archaebacteria (Lorenzo et al. 1999) AIF is ubiquitously expressed, both in normal tissues and in a variety of cancer cell lines. The AIF precursor is synthesized in the cytosol and is imported into mitochondria.

The mature AIF protein, a flavoprotein (prosthetic group, FAD) with significant homology to plant ascorbate reductases and bacterial NADH oxidases, is normally confined to the mitochondrial intermembrane space (Susin et al. 1999; Daugas et al. 2000). Accordingly, transient tranfection of COS cells with GFP fused to the C-terminus of AIF to generate a chimeric AIF-GFP protein targets GFP to mitochondria (Loeffler et al. 2001). When cells are induced to undergo apoptosis, AIF is released from mitochondrial to the cytosol and translocates to the nucleus. This process has been confirmed using all available methods for the determination of the subcellular localization of AIF: in situ immunostaining (Susin et al. 1999; Daugas et al. 2000; Ferri et al. 2000b), transfection with AIF-GFP (Loeffler et al. 2001), and subcellular fractionation (Susin et al. 1999; Ferri et al. 2000b).

When added to purified nuclei, AIF induces peripheral nuclear chromatin condensation, as well as large scale ($\sim$50 kbp) DNA fragmentation. If mi-

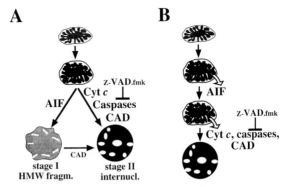

Fig. 1 A, B. Two alternative models of apoptosis regulation by AIF. **A** Alternative action of the caspase-dependent and AIF-dependent pathways linking mitochondrial membrane permeabilization (MMP) to nuclear apoptosis. Mitochondria release AIF and cytochrome c (Cyt c) in a near-to-simultaneous fashion as a consquence of MMP. AIF would be responsible for the peripheral chromatin condensation (stage I) and high molecular weight (HMW) DNA fragmenation observed in cells in which the cascade culminating in the activation of caspase-activated DNAse (CAD) is prevented, for instance by addition of the pancaspase inhibitor Z-VAD.fmk. CAD would be responsible for the advanced chromatin condensation with nuclear shrinkage and formation of nuclear bodies (stage II) as well as the oligonucleosomal "ladder-type" DNA fragmentation in the nucleus. **B** Sequential action of AIF and caspases. In this model, AIF would be released from mitochondria before the caspase co-activator Cyt c. This early release of AIF would be necessary for the subsequent Cyt c release and caspase activation culminating in nuclear apoptosis. Note that in **A** the nuclear effects of extramitochondrial AIF predominate whereas in **B** the cytoplasmic effects are more prominent. Experimental evidence has been obtained in favor of the two models in different paradigms of cell death. The molecular mechanisms accounting for the difference between the two models are elusive

croinjected into the cytoplasm of normal cells, recombinant AIF suffices to cause several hallmarks of apoptosis, including the condensation of nuclear chromatin and the exposure of phosphatidylserine on the plasma membrane surface (Susin et al. 1999; Ferri et al. 2000b; Loeffler et al. 2001). These alterations are rapid (30–120 min) and are not prevented by addition of the pan-caspase inhibitor Z-VAD.fmk (Susin et al. 1999; Ferri et al. 2000b). Moreover, they are not affected by overexpression of the anti-apoptotic protein Bcl-2 (Susin et al. 1999). Similar in vivo effects have been obtained by transfection-enforced overexpression of a truncated AIF-GFP construct misdirected to the extra-mitochondrial compartment due to the deletion of the N-terminal MLS (AIF-GFPΔ1–100; Loeffler et al. 2001).

The intracellular neutralization of extramitochondrial AIF by microinjection of an AIF-specific antibody has established two alternative models of AIF involvement in cell death (Fig. 1). In mouse embryonic fibroblasts, AIF neutralization only ablishes nuclear apoptosis when the apoptosome/caspase-3/CAD pathway is simultaneously blocked by genetic invalidation of Apaf-1 or caspase-3, addition of the pan-caspase inhibitor Z-VAD.fmk, or microinjection of ICAD (Susin et al. 2000). Such data indicate the co-existence of two alternative pathways linking mitochondria to nuclear apoptosis, one that requires AIF and another that relies on caspase activation. In sharp contrast, in other models of cell death (Rat-1 cells treated with staurosporin or HeLa cells dying upon interaction between the HIV-1 envelope glycoprotein complex and CD4/CXCR4), microinjection of AIF antibody blocks the release of cytochrome c and subsequent caspase activation (while cytochrome c neutralization and caspase inhibition do not affect the release of AIF), placing AIF upstream of the cytochrome c/Apaf-l/caspase/CAD pathway (Susin et al. 1999; Ferri et al. 2000 b). Accordingly, when added to purified mitochondria in combination with cytosolic extract, recombinant AIF causes the permeabilization of the outer mitochondrial membrane, leading to the release of cytochrome c and subsequent caspase activation (Susin et al. 1999). Moreover, transfection-enforced overexpression of AIF in Rat-1 cells (Loeffler et al. 2001) or microinjection of recombinant AIF into HeLa syncytia (Ferri et al. 2000 b) can trigger the release of cytochrome c from mitochondria. As a result, AIF may act as a facultative signalling molecule upstream (or at the level) of mitochondria, as well as a death effector downstream of mitochondria.

Open Questions and Perspectives

The mitochondrion has recently emerged as a major player in the regulation of apoptosis, including in neurons, where it may play a prime role in the processes of ischemia/reperfusion damage, excitotoxicity, and neurodegeneration. However, it is a matter of debate how signals affecting mitochondria or primary alterations in mitochondrial metabolism relay to MMP and whether pharmacological interventions on MMP may have a prolonged neuroprotective effect in patients, either in acute or in chronic processes leading to neuronal cell death. At present, it is also unclear to what extent caspase-dependent and caspase-independent (AIF-dependent?) pathways are regulating pathological neuronal cell death. Future investigation involving genetic inactivation of AIF in the mouse may furnisch important insights into the pathophysiology of neuronal cell death.

Acknowledgments. This work has been supported by a special grant from the Ligue Nationale contre le Cancer, Comité Val de Marne de la Ligue contre le Cancer, as well as grants from ANRS, FRM, and the European Commission (to G.K.). L.R. receives a fellowship from the French Ministry for Science and Technology.

References

Andreyev A, Fiskum G (1999) Calcium induced release of mitochondrial cytochrome c by different mechanisms selective for brain versus liver. Cell Death Differ 6:825–832

Berman SB, Hastings TG (1999) Dopamine oxidation alters mitochondrial respiration and induces permeability transition in brain mitochondria: Implication for Parkinson's disease. J Neurochem 73:1127–1137

Berman SB, Watkins SC, Hastings TG (2000) Quantitative biochemical and ultrastructural comparison of mitochondrial permeability transition in isolated brain and liver mitochondria: evidence for reduced sensitivity of brain mitochondria. Exp Neurol 164:415–425

Bernardi P, Scorrano L, Colonna R, Petronilli V, Di Lisa F (1999) Mitochondria and cell death – Mechanistic aspects and methodological issues. Eur J Biochem 264:687–701

Brenner C, Cardiou H, Vieira HLA, Zamzami N, Marzo I, Xie Z, Leber B, Andrews D, Duclohier H, Reed JC, Kroemer G (2000) Bcl-2 and Bax regulate the channel activity of the mitochondrial adenine nucleotide translocator. Oncogene 19:329–336

Brouillet E, Conde F, Beal MF, Hantraye P (1999) Replicating Huntington's disease in experimental animals. Prog Neurobiol 59:427–468

Budijardjo I, Oliver H, Lutter M, Luo X, Wang X (1999) Biochemical pathways of caspase activation during apoptosis. Annu Rev Cell Dev Biol 15:269–290

Carri MT, Ferri A, Battistoni A, Famhy L, Gabbianelli R, Poccia F, Rotilio F (1997) Expression of Cu, Zn superoxide dismutase typical of familial amyotrophic lateral sclerosis induces mitochondrial alteration and increase of cytosolic Ca^{2+} concentration in transfected neuroblastoma SH-SY5Y cells. FEBS Lett 414:365–368

Casari G, De Fusco M, Ciarmatori S, Zeviani M, Mora M, Fernandez P, De Michele G, Filla A, Cocozza S, Marconi R, Dürr A, Fontaine B, Ballabio A (1998) Spastic paraplegia and OXPHOS impairment caused by mutations in paraplegin, a nuclear-encoded mitochondrial metalloprotease. Cell 93:973–983

Cassarino DS, Swerdlow RH, Parks JK, Parker WD, Bennett JP (1998) Cyclosporin A increases resting mitochondrial membrane potential in SY5Y cells and reverses the depressed mitochondrial membrane potential of Alzheimer's disease cybrids. Biochem Biopys Res Commun 248:168–173

Cassarino DS, Pars JK, Parker WD, Bennett JP (1999) The parkinsonian neurotoxin MPP+ opens the mitochondrial permeability transition pore and releases cytochrome c in isolated mitochondria via an oxidative mechanism. Biochim Biophys Acta 1453:49–62

Cooper JM, Schapira AHV (1997) Mitochondrial dysfunction in neurodegeneration. J Bioenerg Biomembr 29:175–183

Costantini P, Jacotot E, Decaudin D, Kroemer G (2000) The mitochondrion as a novel target of anti-cancer chemotherapy. J Natl Cancer Inst, in press

Crompton M (1999) The mitochondrial permeability transition pore and its role in cell death. Biochem J 341:233–249

Crompton M, Virji S, Ward JM (1998) Cyclophilin-D binds strongly to complexes of the voltage-dependent anion channel and the adenine nucleotide translocase to form the permeability transition pore. Eur J Biochem 258:729–735

Daugas E, Susin SA, Zamzami N, Ferri K, Irinopoulos T, Larochette N, Prevost MC, Leber B, Andrews D, Penniger J, Kroemer G (2000) Mitochondrio-nuclear redistribution of AIF in apoptosis and necrosis. FASEB J 14:729–739

de la Monte SM, Luong T, Neely TR, Robinson D, Wands JR (2000) Mitochondrial DNA damage as a mechanism of cell loss in Alsheimer's disease. Lab Invest 80:1323–1335

Desagher S, Osen-Sand A, Nichols A, Eskes R, Montessuit S, Lauper S, Maundrell K, Antonsson B, Martinou J-C (1999) Bid-induced conformational change of Bax is responsible for mitochondrial cytochrome c release during appoptosis. J Cell Biol 144:891–901

Eskes R, Desagher S, Antonsson B, Martinou JC (2000) Bid induces the oligomerization and insertion of Bax into the outer mitochondrial membrane. Mol Cell Biol 20:929–935

Ferrand Drake M, Friberg H, Wieloch T (1999) Mitochondrial permeability transition induced DNA-fragmentation in the rat hippocampus following hypoglycemia. Neuroscience 90:1325–1338

Ferrante RJ, Andreassen OA, Jenkins BG, Dedeoglu A, Kuemmerle S, Kubilus JK, Kaddurah-Daouk R, Hersch SM, Beal MF (2000) Neuroprotective effects of creatine in a transgenic mouse model of Huntington's disease. J Neurosci 20:4389–4397

Ferri FK, Jacotot E, Blanco J, Esté JA, Kroemer G (2000a) Mitochondrial control of cell death induced by proteins encoded by HIV-1. Ann NY Acad Sci 926:149–164

Ferri KF, Jacotot E, Blanco J, Esté JA, Zamzami A, Susin SA, Brothers G, Reed JC, Penninger JM, Kroemer G (2000b) Apoptosis control in syncytia induced by the HIV-1-envelope glycoprotein complex. Role of mitochondria and caspases. J Exp Med 192:1081–1092

Fontaine E, Ichas F, Bernardi P (1998) A ubiquinone-binding site regulates the mitochondrial permeability transition pore. J Biol Chem 273:25734–25740

Goldstein JC, Waterhouse NJ, Juin P, Evan GI, Green DR (2000) The coordinate release of cytochrome c is rapid, complete and kinetically invariant. Nat Cell Biol 2:156–162

Green DR, Kroemer G (1998) The central executioner of apoptosis: mitochondria or caspase? Trends Cell Biol 8:267–271

Green DR, Reed JC (1998) Mitochondria and apoptosis. Science 281:1309–1312

Gross A, McDonnell JM, Korsmeyer SJ (1999) Bcl-2 family members and the mitochondria in apoptosis. Genes Dev 13:1988–1911

Guo Q, Fu WM, Holtsberg FW, Steiner SM, Mattson MP (1999) Superoxide mediates the cell-death-enhancing action of presenilin-1 mutations. J Neurosci Res 56:457–470

Heiskanen KM, Bhat MB, Wang HW, Ma JJ, Nieminen AL (1999) Mitochondrial depopolarization accompanies cytochrome c release during apoptosis in PC6 cells. J Biol Chem 274:5654–5658

Jacotot E, Ravagnan L, Loeffler M, Ferri KF, Vieira HLA, Zamzami N, Costantini P, Druillennec S, Hoebeke J, Brian JP, Irinopoulos T, Daugas E, Susin SA, Cointe D, Xie ZH, Reed JC, Roques BP, Kroemer G (2000) The HIV-1 viral protein R induces apoptosis via a direct effect on the mitochondrial permeability transition pore. J Exp Med 191:33–45

Knaspekov L, Friberg H, Halestrap A, Viktorov I, Wieloch T (1999) Cyclosporin A and its nonimmunosuppressive analogue N-Me-Val-4-cylosporin A mitigate glucose/oxygen deprivation-induced damage to rat cultured hippocampal neurons. Eur J Neurosci 11:3194–3198

Klivenyi P, Ferrante RJ, Matthews RT, Bogdnaow MB, Klein AL, Andrassen OA, Mueller G, Wermer M, Kaddurah-Daouk R, Beal F (1999) Neuroprotective effects of creatine in a transgenic animal model of amyotrophic lateral sclerosis. Nat Med 5:347–350

Kostic V, Jackson-Lewis V, de Bilbao F, Dubois-Dauphin M, Przedborski S (1997) Bcl-2: Prolonging life in a transgenic mouse model of familial amyotrophic lateral sclerosis. Science 277:559–562

Koutnikova H, Campuzano V, Foury F, Dollé P, Cazzalini O, Koenig M (1997) Studies of human, mouse, and yeast homologues indicate a mitochondrial function for frataxin. Nat Genet 16:345–351

Kroemer G (1997) The proto-oncogene Bcl-2 and its role in regulating apoptosis. Nat Med 3:614–620

Kroemer G, Reed JC (2000) Mitochondrial control of cell death. Nat Med 6:513–519

Kroemer G, Dallaporta B, Resche-Rigon M (1998) The mitochondrial death/life regulator in apoptosis and necrosis. Annu Rev Physiol 60:619–642

Kwong J, Choi HL, Huang Y, Chan FL (1999) Ultrastructural and biochemical observations on the early changes in apoptotic epithelial cells of the rat prostate induced by castration. Cell Tissue Res 298:123–136

Lemasters JJ, Nieminen A-L, Qjan T, Trost LC, Elmore SP, Nishimura Y, Crowe RA, Cascio WE, Bradham CA, Brenner DA, Herman B (1998) The mitochondrial permeability transition in cell death: a common mechanism in necrosis, apoptosis and autophagy. Biochim Biophys Acta 1366:177–196

Leventhal L, Sortwell CE, Hanbury R, Collier TJ, Kordower JH, Palfi S (2000) Cyclosporin A protects striatal neurons in vitro and in vivo from 3-nitroproprionic acid toxicity. J Comp Neurol 425:471–478

Li H, Kolluri SK, Gu J, Daxson MJ, Cao X, Hobbs PD, Lin B, Chen G-Q, Lu J-S, Lin F, Xie Z, Fontana JA, Reed JC, Zhang X-K (2000) Cytochrome c release and apoptosis induced by mitochondrial targeting of orphan receptor TR3-nur77-NGFI-B. Science 289:1159–1164

Liu XS, Kim CN, Yang J, Jemmerson R, Wang X (1996) Induction of apoptotic program in cell-free extracts: requirement for dATP and cytochrome C. Cell 86:147–157

Loeffler M, Daugas E, Susin SA, Zamzami N, Métivier D, Nieminen AL, Brothers G, Penninger JM, Kroemer G (2001) Dominant cell death induction by extramitochondrially targeted apoptosis-inducing factor. FASEB J 15:758–767

Lorenzo HK, Susin SA, Penninger J, Kroemer G (1999) Apoptosis inducing factor (AIF): a phylogenetically old, caspase-independent effector of cell death. Cell Death Differ 6:516–524

Marchenko ND, Zaika A, Moll UM (2000) Death signal-induced localization of p53 protein to mitochondria. A potential role in apoptotic signaling. J Biol Chem 275:16202–16212

Marchetti P, Castedo M, Susin SA, Zamzami N, Hirsch T, Haeffner A, Hirsch F, Geuskens M, Kroemer G (1996) Mitochondrial permeability transition is a central coordinating event of apoptosis. J Exp Med 184:1155–1160

Marzo I, Brenner C, Zamzami N, Jürgensmeier J, Susin SA, Vieira HLA, Prévost M-C, Xie Z, Mutsiyama S, Reed JC, Kroemer G (1998a) Bax and adenine nucleotide translocator cooperate in the mitochondrial control of apoptosis. Science 281:2027–2031

Marzo I, Brenner C, Zamzami N, Susin SA, Beutner G, Brdiczka D, Rémy R, Xie Z-H, Reed JC, Kroemer G (1998b) The permeability transition pore complex: a target for apoptosis regulation by caspases and Bcl-2 related proteins. J Exp Med 187:1261–1271

Matthews RT, Yang LC, Browne S, Baik M, Beal MF (1998) Coenzyme Q(10) administration increases brain mitochondrial concentrations and exerts neuroprotective effects. Proc Natl Acad Sci USA 95:8892–8897

Okonkwo DO, Buki A, Siman R, Povlishock JT (1999) Cyclosporin A limits calcium-induced axonal damage following traumatic brain injury. Neuroreport 10:353–358

Passer BJ, Pellegrini L, Vito P, Ganjei JK, D'Adamio L (1999) Interaction of Alzheimer's presenilin-1 and presenilin-2 with Bcl-XL-A potential role in modulating the threshold of cell death. J Biol Chem 274:24007–24013

Pastorino JG, Tafani M, Rothman RJ, Macineviciute A, Hoek JB, Farber JL (1999) Functional consequences of sustained or transient activation by Bax of the mitochondrial permeability transition pore. J Biol Chem 274:31734–31739

Patterson S, Spahr CS, Daugas E, Susin SA, Irinopoulos T, Koehler C, Kroemer G (2000) Mass spectrometric identification of proteins released from mitochondria undergoing permeability transition. Cell Death Differ 7:137–144

Petersen A, Castilho RF, Hansson O, Wieloch T, Brundin P (2000) Oxidative stress, mitochondrial permeability transition and activation of caspases in calcium ionophore A23187-induced death of cultured striatal neurons. Brain Res 857:20–29

Petronilli V, Miotto G, Canton M, Brini M, Colonna R, Bernardi P, Di Lisa F (1999) Transient and long-lasting openings of the mitochondrial permeability transition pore can be monitored directly in intact cells by changes in mitochondrial calcein fluorescence. Biophys J 76:725–734

Ravagnan L, Marzo I, Costantini P, Susin SA, Zamzami N, Petit PX, Hirsch F, Poupon M-F, Miccoli L, Xie Z, Reed JC, Kroemer G (1999) Lonidamine triggers apoptosis via a direct, Bcl-2-inhibited effect on the mitochondrial permeability transition pore. Oncogene 18:2537–2546

Saito M, Korsmeyer SJ, Schlesinger PH (2000) Bax-dependent transport of cytochrome c reconstituted in pure liposomes. Nat Cell Biol 2:553–555

Sawa A, Wiegand GW, Cooper J, Margolis RL, Sharp AH, Lawler JF, Greenamyre JT, Synder SH, Ross CA (1999) Increased apoptosis of Huntingon disease lymphoblasts associated with repeat length-dependent mitochondrial depolarization. Nat Med 5:1194–1198

Sheehan JP, Swerdlow RH, Parker WD, Miller SW, Davis RE, Tuttle JB (1997) Altered calcium homeostasis in cells transformed by mitochondria from individuals with Parkinson disease. J Neurochem 68:1221–1233

Shimizu S, Narita M, Tsujimoto Y (1999) Bcl-2 family proteins regulate the release of apoptogenic cytochrome c by the mitochondrial channel VDAC. Nature 399:483–487

Shimizu S, Ide T, Yanagida T, Tsujimoti Y (2000) Electrophysiological study of a novel large pore formed by Bax and the voltage-dependent anion channel that is premeable to cytochrome c. J Biol Chem 275:12321–12325

Susin SA, Zamzami N, Castedo M, Hirsch T, Marchetti P, Macho A, Daugas E, Geuskens M, Kroemer G (1996) Bcl-2 inhibits the mitochondrial release of an apoptogenic protease. J Exp Med 184:1331–1342

Susin SA, Lorenzo HK, Zamzami N, Marzo I, Snow BE, Brothers GM, Mangion J, Jacotot E, Costantini P, Loeffler M, Larochette N, Goodlett DR, Aebersold R, Siderovski DP, Penninger JM, Kroemer G (1999) Molecular characterization of mitochondrial apoptosis-inducing factor. Nature 397:441–446

Susin SA, Daugas E, Ravagnan L, Samejima K, Zamzami N, Loeffler M, Costantini P, Ferri KF, Irinopoulou T, Prévost M-C, Brothers G, Mak TW, Penninger J, Earnshaw WC, Kroemer G (2000) Two distinct pathways leading to nuclear apoptosis. J Exp Med 192:577–585

Szalai G, Kirschnamurthy R, Hajnoczky G (1999) Apoptosis driven by IP3-linked mitochondrial calcium signals. EMBO J 18:6349–6361

Van der Heiden M, Chandel NS, Schumacker PT, Thompson CB (1999) Bcl-XL prevents cell death following growth factor withdrawal by facilitating mitochondrial ATP/ADP exchange. Mol Cell 3:159–167

Vander Heiden MG, Thompson CB (1999) Bcl-2 proteins: Inhibitors of apoptosis or regulators of mitochondrial homeostasis? Nat Cell Biol 1:E209–E216

Wang HG, Pathan N, Ethell IM, Krajewski S, Yamaguchi Y, Shibasaki F, McKean F, Bobo T, Franke TF, Reed JC (1999) Ca2+-induced apoptosis through calcineurin dephosphorylation of Bad. Science 284:339–343

Wei MC, Lindsten T, Mootha VK, Weiler S, Gross A, Ashiya M, Thompson CB, Korsmeyer SJ (2000) tBID, a membrane-targeted death ligand, oligomerizes BAK to release cytochrome c. Genes Dev 14:2060–2071

Yoshimoto T, Siesjo BK (1999) Posttreatment with the immunosuppressant cyclosporin A in transient focal ischemia. Brain Res 839:283–291

Zamzami N, Marchetti P, Castedo M, Decaudin D, Macho A, Hirsch T, Susin SA, Petit PX, Mignotte B, Kroemer G (1995a) Sequential reduction of mitochondrial transmembrane potential and generation of reactive oxygen species in early programmed cell death. J Exp Med 182:367–377

Zamzami N, Marchetti P, Castedo M, Zanin C, Vayssière J-L, Petit PX, Kroemer G (1995b) Reduction in mitochondrial potential constitutes an early irreversible step of programmed lymphocyte death in vivo. J Exp Med 181:1661–1672

Zamzami N, Susin SA, Marchetti P, Hirsch T, Gómex-Monterrey I, Castedo M, Kroemer G (1996) Mitochondrial control of nuclear apoptosis. J Exp Med 183:1533–1544

Bcl-2 Family Members and Permeabilization of the Outer Mitochondrial Membrane

O. Terradillos, X. Roucou, S. Da Cruz, B. Sanchez, and J.-C. Martinou

Summary

Mitochondria play a major role in many apoptotic responses. They coordinate caspase activation through the release of cytochrome c and DIABLO/Smac as a result of the outer mitochondrial membrane becoming permeable. The release of these proteins is controlled by Bcl-2 family members. The mechanisms whereby Bcl-2 family members control the permeability of the outer mitochondrial membrane are still unclear.

Apoptosis is an essential process required for the development and maintenance of tissue homeostasis. Increased apoptosis could contribute to neurodegenerative diseases. Central components of the death machinery include the Bcl-2, apoptotic protease-activating factor-1 (Apaf-1) and caspase family members. The caspases are cysteine proteases that cleave key intracellular substrates, resulting in the morphological and biochemical changes associated with apoptosis. Recently, it was found that, in many apoptotic responses, caspase activation is coordinated by mitochondria.

Mitochondria as the Coordinator of Caspase Activation

Mitochondria coordinate caspase activation through the release into the cytosol of apoptogenic factors such as cytochrome c and Smac/DIABLO (Du et al. 2000; Verhagen et al. 2000; Desagher and Martinou 2000). The current model is that, once in the cytosol, cytochrome c binds to Apaf-1, which, in the presence of ATP, recruits and activates pro-caspase-9. Activated caspase-9 can in turn activate other caspases in charge of cell execution (Budihardjo et al. 1999). As to Smac/DIABLO, once released into the cytosol, it binds to inhibitors of apoptosis (IAPs) that are natural caspase inhibitors, thereby contributing to caspase activation. In addition to cytochrome c and Smac/DIABLO, many other proteins normally confined to the intermembrane space of mitochondria are released during apoptosis. These include apoptosis inducing factor (AIF; Susin et al. 1999), adenylate kinase and sulfite oxidase (Kluck et al. 1999; Samali et al. 1999; Priault et al. 1999). So far, only cytochrome c appears essential for apoptosis triggered in neurons (Neame et al. 1998) and fibroblasts (Juin et al. 1999) by growth factor deprivation or c-myc

Henderson/Green/Mariani/Christen (Eds.)
Neuronal Death by Accident or by Design
© Springer-Verlag Berlin Heidelberg 2001

expression, respectively. Indeed, neutralization of cytochrome c with antibodies injected into the cytosol is sufficient to protect these cells. The release of mitochondrial proteins is controlled by the Bcl-2 family members.

Mitochondria as a Major Target for Bcl-2 Family Proteins

Until now, 15 Bcl-2 family proteins have been identified in mammals (Kroemer 1997; Gross et al. 1999; Adams and Cory 1998); they can be subdivided into anti- (e.g., Bcl-2 and Bcl-x_L) and pro-apoptotic (e.g., Bax and Bak) members. Many of them are targeted to the mitochondrial outer membrane through a C terminal hydrophobic domain, and the principal mechanism by which these proteins regulate apoptosis is probably through the control of cytochrome c release. The anti-apoptotic proteins such as Bcl-2 or Bcl-x_L prevent cytochrome c release from mitochondria and caspase activation, whereas pro-apoptotic proteins such as Bax or Bak promote cytochrome c release from mitochondria. Caspase inhibitors do not affect Bax-induced cytochrome c leakage, whereas they effectively block caspase activation and delay apoptosis (Rossé et al. 1998; Jürgensmeier et al. 1998; Eskes et al. 1998; Finucane et al. 1999). These data suggest that Bax can directly induce loss of cytochrome c and indicate that caspases are activated downstream of cytochrome c release.

Although Bcl-2 appears to be exclusively membrane bound, in particular in mitochondria (Hsu et al. 1997), other proteins such as Bid, Bad and Bim are cytosolic but translocate to mitochondria during apoptosis (Desagher et al. 1999; Zha et al. 1996; Puthalalath et al. 1999). These proteins play a major role in transducing signals from the cytosol to mitochondria, where they bind to and regulate the activity of the Bcl-2 family members that control the release of cytochrome c. Translocation of these proteins is triggered by specific post-translational modifications such as, for example, phosphorylation in the case of Bad (Zha et al. 1996; Datta et al. 1997; del Peso et al. 1997; Scheid et al. 1999; Harada et al. 1999) or cleavage by caspase-8 in the case of Bid (Li et al. 1998; Luo et al. 1998). Likewise, in healthy cells, Bax appears to be predominantly cytosolic but it can also be associated with mitochondria. Following a death signal, several changes affect Bax which lead to its activation. Many data suggest that, first, Bax undergoes a change of conformation that unmasks its N-terminal part, as indicated by reactivity to epitope-specific antibodies and by sensitivity to exogenous proteolysis (Desagher et al. 1999; Goping et al. 1998) and probably its C-terminus as well, which triggers its translocation to mitochondria (Hsu et al. 1997; Gross et al. 1998). Then, Bax becomes an integral membrane protein (Goping et al. 1998; Eskes et al. 2000) and oligomerizes. The mechanisms responsible for its oligomerization (Gross et al. 1998; Eskes et al. 2000), which occurs in the outer mitochondrial membrane are still unclear. These Bax changes can be prevented by Bcl-2 and Bcl-x_L (Desagher et al. 1999; Gross et al. 1998) or by E1B19K. Similar data have been reported for Bak, except that, in contrast to

Bax, Bak appears to be constitutively localized in mitochondria. Oligomerization of Bax or Bak is followed by permeabilization of the outer mitochondrial membrane and the release of cytochrome c and other proteins from the intermembrane space of mitochondria.

How is Cytochrome c Released from Mitochondria?

Several competing models have been proposed to explain how cytochrome c is released from mitochondria during apoptosis. Two models share the common prediction that the outer mitochondrial membrane ruptures as a result of mitochondrial matrix swelling. The first postulates the opening of a megachannel called the permeability transition pore (PTP). This channel is poorly characterized in molecular terms but is proposed to span both mitochondrial membranes at sites where the two membranes are apposed. The adenine nucleotide translocator (ANT, in the inner mitochondrial membrane) and the voltage-dependent anion channel (VDAC, in the outer mitochondrial membrane) are considered to be major components of the PTP. According to the PTP model, PTP openers, including Bax, cause inner membrane permeabilization and mitochondrial depolarization via binding to the ANT (Marzo et al. 1998). This process allows the entry of water and solutes into the matrix and leads to mitochondrial swelling. The other model postulates a defect in mitochondrial ATP/ADP exchange owing to VDAC closure (Vander Heiden and Thompson 1999). This leads to a hyperpolarization of the inner mitochondrial membrane and subsequent matrix swelling. Both models may account for some but not all types of cell death. For example, they do not explain why in certain conditions the drop in membrane potential or the changes in mitochondrial morphology appear to be late events that follow the release of mitochondrial proteins.

Two other models do not predict damage of the outer mitochondrial membrane but rather the formation of a pore in the outer mitochondrial membrane that is large enough to allow the passage of cytochrome c and other mitochondrial proteins of the intermembrane space into the cytosol. A likely candidate for the formation of this pore is Bax (or a Bax relative). Indeed, an oligomer of Bax can form large conductance channels in lipid planar bilayers (Schlesinger et al. 1997; Antonsson et al. 1997). Moreover, addition of Bax directly to isolated mitochondria triggers cytochrome c release through a mechanism insensitive to PTP blockers (Eskes et al. 1998) and does not involve mitochondrial swelling (Jürgensmeier et al. 1998). Bax also causes membrane instability by decreasing the linear tension of phospholipid bilayers (Basañez et al. 1999). This effect could facilitate its insertion and oligomerization within membranes, leading to formation of lipidic pores or lipid/protein complexes. Yet another model involves Bax cooperating with VDAC to form a cytochrome c conducting channel (Shimizu et al. 1999). Following its binding with Bax, the conformation of VDAC would change, leading to the formation of a channel permeable to cytochrome c. Although a theory involving a pore in the outer mitochondrial membrane looks attractive, no one has yet seen such a structure in mitochondria during the process of apoptosis. The search for Bax channels goes on.

Concluding Remarks

Permeabilization of mitochondrial membranes appears to be a common event for many forms of cell death. However, there are apoptotic responses in which mitochondria play only a minor role. For example, apoptosis triggered by death receptors of the TNF/Fas family does not depend on mitochondria except for cells in which caspase-8, which is engaged by death receptors, is unable to activate downstream caspases. In this case, mitochondria are connected to the pathway triggered by death ligands through the cleavage of Bid.

The mechanisms that lead to permeabilization of the outer mitochondrial membrane may result from diverse mechanisms depending on the death stimulus and the cell type in which this process occurs. Understanding these mechanisms more precisely may allow the identification of new therapeutic targets for the development of drugs that will modulate cell death associated with many pathological states.

References

Adams JM, Cory S (1998) The Bcl-2 protein family: arbiters of cell survival. Science 281:1322–1326

Antonsson B, Conti F, Ciavatta A, Montessuit S, Lewis S, Martinou I, Bernasconi L, Bernard A, Mermod JJ, Mazzei G, Maundrell K, Gambale F, Sadoul R, Martinou JC (1997) Inhibition of Bax channel-forming activity by Bcl-2. Science 277:370–372

Basañez G, Nechushtan A, Drozhinin O, Chanturiya A, Choe E, Tutt S, Wood KA, Hsu Y-T, Zimmerberg J, Youle RJ (1999) Bax, but not Bcl-x$_L$, decreases the lifetime of planar phospholipid bilayer membranes at subnanomolar concentrations. Proc Natl Acad Sci USA 96:5492–5497

Budihardjo I, Oliver H, Lutter M, Luo X, Wang X (1999) Biochemical pathways of caspase activation during apoptosis. Ann Rev Cell Dev Biol 15:269–290

Datta SR, Dudek H, Tao X, Masters S, Fu H, Gotoh Y, Greenberg ME (1997) Akt phosphorylation of BAD couples survival signals to the cell-intrinsic death machinery. Cell 91:231–241

del Peso L, González-García M, Page C, Herrera R, Nuñez G (1997) Interleukin-3-induced phosphorylation of BAD through the protein kinase Akt. Science 278:687–689

Desagher S, Martinou J-C (2000) Mitochondria as the central control point of apoptosis. Trends Cell Biol 10:369–377

Desagher S, Osen-Sand A, Nichols A, Eskes R, Montessuit S, Lauper S, Maundrell K, Antonsson B, Martinou JC (1999) Bid-induced conformational change of Bax is responsible for mitochondrial cytochrome c release during apoptosis. J Cell Biol 144:891–901

Du C, Fang M, Li Y, Li L, Wang X (2000) Smac, a mitochondrial protein that promotes cytochrome c-dependent caspase activation by eliminating IAP inhibition. Cell 102:33–42

Eskes R, Antonsson B, Osen-Sand A, Montessuit S, Richter C, Sadoul R, Mazze G, Nichols A, Martinou JC (1998) Bax-induced cytochrome C release from mitochondria is independent of the permeability transition pore but highly dependent on Mg^{2+} ions. J Cell Biol 143:217–224

Eskes R, Desagher S, Antonsson B, Martinou J-C (2000) Bid induces the oligomerization and insertion of Bax into the outer mitochondrial membrane. Mol Cell Biol 20:929–935

Finucane DM, Bossy-Wetzel E, Waterhouse NJ, Cotter TG, Green DR (1999) Bax-induced caspase activation and apoptosis via cytochrome c release from mitochondria is inhibitable by Bcl-x$_L$. J Biol Chem 274:2225–2233

Goping IS, Gross A, Lavoie JN, Nguyen M, Jemmerson R, Roth K, Korsmeyer SJ, Shore GC (1998) Regulated targeting of BAX to mitochondria. J Cell Biol 143:207–215

Gross A, Jockel J, Wei MC, Korsmeyer SJ (1998) Enforced dimerization of BAX results in its translocation, mitochondrial dysfunction and apoptosis. EMBO J 17:3878–3885

Gross A, McDonnell JM, Korsmeyer SJ (1999) BCL-2 family members and the mitochondria in apoptosis. Genes Devel 13:1899–1911

Harada H, Becknell B, Wilm M, Mann M, Huang LJ, Taylor SS, Scott JD, Korsmeyer SJ (1999) Phosphorylation and inactivation of BAD by mitochondria-anchored protein kinase A. Mol Cell 3:413–422

Hsu Y-T, Wolter KG, Youle RJ (1997) Cytosol-to-membrane redistribution of Bax and Bcl-X$_L$ during apoptosis. Proc Natl Acad Sci USA 94:3668–3672

Juin P, Hueber AO, Littlewood T, Evan G (1999) c-Myc-induced sensitization to apoptosis is mediated through cytochrome c. Genes Devel 13:1367–1381

Jürgensmeier JM, Xie Z, Deveraux Q, Ellerby L, Bredesen D, Reed JC (1998) Bax directly induces release of cytochrome c from isolated mitochondria. Proc Natl Acad Sci USA 95:4997–5002

Kluck RM, Esposti MD, Perkins G, Renken C, Kuwana T, Bossy-Wetzel E, Goldberg M, Allen T, Barber MJ, Green DR, Newmeyer DD (1999) The pro-apoptotic proteins, Bid and Bax, cause a limited permeabilization of the mitochondrial outer membrane that is enhanced by cytosol. J Cell Biol 147:809–822

Kroemer GC (1997) The proto-oncogene Bcl-2 and its role in regulating apoptosis. Nature Med 3:614–620

Li H, Zhu H, Xu C, Yuan J (1998) Cleavage of BID by Caspase 8 mediates the mitochondrial damage in the Fas pathway of apoptosis. Cell 94:491–501

Luo X, Budihardjo I, Zou H, Slaughter C, Wang X (1998) Bid, a Bcl-2 interacting protein, mediates cytochrome c release in response to activation of cell surface death receptors. Cell 94:481–490

Marzo I, Brenner C, Zamzami N, Jürgensmeier JM, Susin SA, Vieira HL, Prevost MC, Xie Z, Matsuyama S, Reed JC, Kroemer G (1998) Bax and adenine nucleotide translocator cooperate in the mitochondrial control of apoptosis. Science 281:2027–2031

Neame SJ, Rubin LL, Philpott KL (1998) Blocking cytochrome c activity within intact neurons inhibits apoptosis. J Cell Biol 142:1583–1593

Priault M, Chaudhuri B, Clow A, Camougrand N, Manon S (1999) Investigation of bax-induced release of cytochrome c from yeast mitochondria. Permeability of mitochondrial membranes, role of VDAC and ATP requirement. Eur J Biochem 260:684–691

Puthalalath G, Huang DCS, O'Reilly LA, King SM, Strasser A (1999) The proapoptotic activity of the Bcl-2 family member Bim is regulated by interaction with the dynein motor complex. Cell 3:287–296

Rossé T, Olivier R, Monney L, Rager M, Conus S, Fellay I, Jansen B, Borner C (1998) Bcl-2 prolongs cell survival after Bax-induced release of cytochrome c. Nature 391:496–499

Samali A, Cai J, Zhivotovsky B, Jones D-P, Orrenius S (1999) Presence of a pre-apoptotic complex of pro-caspase-3, Hsp60 and Hsp-10 in the mitochondrial fraction of Jurkat cells. EMBO J 18:2040–2048

Scheid MP, Schubert KM, Duronio V (1999) Regulation of Bad phosphorylation and association with Bcl-x$_L$ by the MAPK/Erk kinase. J Biol Chem 274:31108–31113

Schlesinger PH, Gross A, Yin XM, Xamamoto K, Saito M, Waksman G, Korsmeyer SJ (1997) Comparison of the ion channel characteristics of proapoptotic BAX and antiapoptotic BCL-2. Proc Natl Acad Sci USA 94:11357–11362

Shimizu S, Narita M, Tsujimoto Y (1999) Bcl-2 family proteins regulate the release of apoptogenic cytochrome c by the mitochondrial channel VDAC. Nature 399:483–487

Susin SA, Lorenzo HK, Zamzami N, Marzo I, Snow BE, Brothers GM, Mangion J, Jacotot E, Costantini P, Loeffler M, Larochette N, Goodlett DR, Aebersold R, Siderovski DP, Penninger JM, Kroemer G (1999) Molecular characterization of mitochondrial apoptosis-inducing factor. Nature 397:441–446

Vander Heiden MG, Thompson CB (1999) Bcl-2 proteins: regulators of apoptosis or of mitochondrial homeostasis? Nature Cell Biol 1:E209–E216

Verhagen AM, Ekert PG, Pakusch M, Silke J, Connolly LM, Reid GE, Moritz RL, Simpson RJ, Vaux DL (2000) Identification of DIABLO, a mammalian protein that promotes apoptosis by binding to and antagonizing IAP proteins. Cell 102:43–53

Zha J, Harada H, Yang E, Jockel J, Korsmeyer SJ (1996) Serine phosphorylation of death agonist BAD in response to survival factor results in binding to 14-3-3 not BCL-X$_L$. Cell 87:619–628

Roles for Fas in Programmed Cell Death of Motoneurons

C. Raoul, G. Ugolini, A. Estevez, B. Pettmann, and C. E. Henderson

Summary

Over the last few years the distinction between naturally occurring and abnormal neuronal cell death mechanisms has become less clear. Not only have stress-related pathways been implicated in developmental programmed cell death, but neuronal loss in human neuro-degenerative disease or stroke has been shown to use some of the same mechanisms as classical apoptosis. The Fas receptor is a "killer molecule" most studied in the immune system. We review here some of the findings concerning its roles in neurons, focussing on results from our laboratory and from others concerning spinal motoneurons. There is now evidence that Fas activation is involved in certain forms of neuronal death, and that this process can be tightly controlled. It is too early to determine the precise role of Fas in the nervous system, either during development or in pathological situations. However, the existence of novel signalling pathways upstream and downstream of Fas in neurons suggests that, if Fas activation does contribute to loss of neurons in human patients, Fas-related pathways may provide a promising new set of targets for therapeutical intervention.

Introduction

Neuronal loss during development was the subject of some of the earliest reports in the programmed cell death (PCD) field. The degree of cell loss in different parts of the nervous system varies, but nearly all identified classes of neuron are produced during development in numbers greater than those in the adult, only to be partially eliminated soon afterwards (reviewed in Oppenheim 1991; Pettmann and Henderson 1998), in some cases well before the period of target contact (Yaginuma et al. 1996). Thus PCD, which is involved in development of many tissues outside the nervous system and probably also in unicellular organisms (reviewed in Ameisen 1996; Jacobson et al. 1997), has been maintained to a very general degree in the nervous system.

Nevertheless, the *raison d'être* of neuronal elimination is still far from clear. Loss of neurons during development does not fulfil a purpose as easily rationalized as sculpting digits, removing a tail or eliminating lymphocytes

Henderson/Green/Mariani/Christen (Eds.)
Neuronal Death by Accident or by Design
© Springer-Verlag Berlin Heidelberg 2001

of unwanted specificities (Jacobson et al. 1997). Moreover, reduction of neuronal death is not necessarily life-threatening in vertebrates. For example, knockout mice for *Bax* (Knudson et al. 1995), and transgenic mice over-expressing Bcl-2 (Martinou et al. 1994), both of which have increased numbers of neurons, show normal life spans in the laboratory environment. However, there may be evolutionary pressures linked to the altered functional properties of adult nervous systems containing too many neurons (Rondi-Reig et al. 1999).

Possible explanations for the generality of neuronal cell death during development include: correction of erroneous projections, creation of pathways for axon outgrowth, numerical limitations imposed by mechanisms involving successive cell doubling, transient functions for the eliminated neurons, and probably many more. Although there is experimental support for some of these ideas in individual systems, none seems to be generally applicable. Indeed, it remains to be demonstrated that a single explanation exists. In vertebrates, the most generally accepted idea is that neurons are produced "in excess" in order that they may compete for contacts with their cellular partners and thus adjust their numbers so as to provide sufficient innervation of their targets. To a large extent, therefore, developmental PCD has been considered as the default pathway in the absence of survival signals. This idea lies at the heart of the "neurotrophic hypothesis" discussed below.

In contrast to this way of thinking about developmental neuronal PCD, abnormal neuronal death in response to stresses such as ischemia, excitotoxicity or other disease processes has often been considered to result from active killing, or activation of stress-related pathways. Several recent developments have contributed to blurring this distinction. First, pathological neuronal death has been shown in some cases to require activation of a cell death cascade, some elements of which are common to developmental and pathological cell death. Second, there is increasing evidence for the role of stress-activated protein kinases in naturally occurring cell death (see below, and chapter by Kuan et al. in this volume). Lastly, active signalling through "death receptors" has been shown to occur in immature neurons during normal development and to play a physiological role in their removal. Many of the earliest data on active killing of developing neurons concerned p75^{NTR}, first studied for its role as a co-receptor for the high-affinity neurotrophin receptors of the Trk family. The fact that p75^{NTR} is a member of the same family of transmembrane receptors as the tumor necrosis factor (TNF) receptor and Fas (CD95/Apo-1) also suggested that it may have other roles. This possibility was subsequently demonstrated by pioneering studies in cell lines, cultured oligodendrocytes and, most strikingly, developing retinal neurons in vivo (reviewed in Bredesen and Rabizadeh 1997; Kaplan and Miller 1997; Casaccia-Bonnefil et al. 1999). Nerve growth factor (NGF) is present in the early chick retina and, although TrkA is absent at these stages, some retinal neurons already express p75^{NTR} (Frade et al. 1996). Application of blocking antibodies to NGF in vivo leads to decreased death of the neurons that express p75^{NTR}, suggesting that in this system endogenous NGF plays a pro-apoptotic role. Since retinal neurons are also saved by antibodies to p75^{NTR} that inhibit binding of NGF, the neuronal killing is probably mediated by the low-affinity receptor (Frade et al. 1996). Strikingly for a developmental process, the cellular source of NGF seems to be the microglia that migrate into the retina at early stages (Frade and Barde 1998). p75^{NTR} knockout mice show reduced neuronal death in the retina and in a dorsal population of spinal cord interneurons (Frade and Barde 1999). Similar changes are found in $ngf^{-/-}$ mutant mice, indicating that NGF signalling through p75^{NTR} is used to eliminate some neurons.

These important findings concerning p75^{NTR} stimulated interest in the potential roles of the other members of the TNF receptor family in nervous system development and pathology.

The Fas/Fas Ligand System

Fas is a cell surface receptor that, when engaged by its ligand (FasL), is capable of triggering the death of the cell that expresses it (reviewed in Nagata and Golstein 1995). Like other members of the TNF receptor family, it has one transmembrane domain and an intracellular "death domain." In the case of Fas, it is the bringing together of these death domains and the recruitment of signalling partners that triggers the cell death cascades. Fas trimerization may occur either following binding of trimeric FasL or spontaneously at the cell membrane (reviewed in Golstein 2000). In the classical Fas signalling pathway (see Fig. 1), the cytoplasmic domain of Fas can then bind the adaptor protein FADD/Mort1 (Chinnaiyan et al. 1995). FADD in turn binds pro-caspase-8, which can thereby self-activate (Muzio et al. 1996). Although in some cell types caspase-8 can directly activate downstream caspases without signalling through Bcl-2 family members in the mitochondrion, in neurons this is not likely to be the case, since nearly all neuronal death is blocked in Bax knockout mice (Deckwerth et al. 1996). For this reason, Figure 1 only illustrates the pathway that involves cleavage of BID, a BH3-only member of the Bcl-2 family, and subsequent activation of Bax, leading to cytochrome c release from the mitochondria (Li et al. 1998; Luo et al. 1998; Desagher et al. 1999; Eskes et al. 2000).

Although the "central pathway" triggered by Fas has been the object of by far the most intensive study, several reports have convincingly described other pathways downstream of Fas activation (Fig. 1). In one of these, activation of the tumor suppressor phosphatase PTEN leads to depletion of PIP-3, thereby inhibiting PI3′ kinase and the Akt kinase survival pathway (Stambolic et al. 1998; Di Cristofano et al. 1999; Sun et al. 1999; Di Cristofano and Pandolfi 2000). In other systems, it has been reported that by acting through the receptor-associated protein Daxx, Fas can activate the JNK kinase kinse ASK1 (Chang et al. 1998). ASK1 is also required for TNF-induced activation of JNK and does not require caspase-8 to induce apoptosis (Nishitoh et al. 1998). Thus, several different pathways are required in different conditions

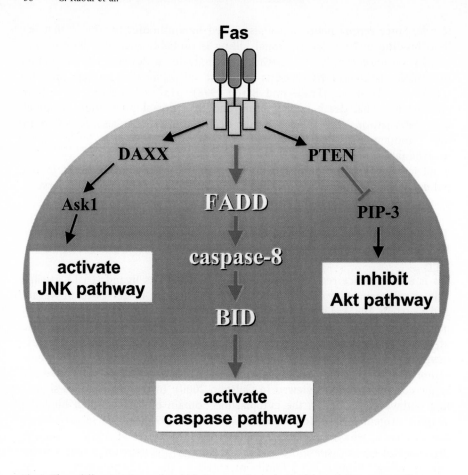

Fig. 1. Three different pathways by which Fas can trigger death of the cell that expresses it

for Fas-triggered PCD. To what extent all these pathways are completely in-hibited in Fas knockout mice (which express truncated Fas without its death domain) has not been determined (Adachi et al. 1995).

The Fas/FasL system has well-studied roles in instructive apoptosis (re-viewed in Nagata 1997). In the immune system, Fas and Fas-L are involved in deletion of mature T cells to end an immune response, in triggering death of inflammatory cells, and in elimination of infected cells or tumors by cyto-toxic T lymphocytes (Nagata and Golstein 1995). FasL can activate Fas both in *trans* and in *cis*: T-lymphocytes induced to die by T-cell receptor ligation up-regulate FasL and Fas at their surface and thus trigger cell death in an autocrine manner (Brunner et al. 1995; Dhein et al. 1995). Partial loss-of-function mutations for Fas (*lpr*) and Fas-L (*gld*) in mouse lead to accumula-tion of peripheral lymphoid cells and to an autoimmune disorder (Wata-nabe-Fukunaga et al. 1992; Takahashi et al. 1994).

The role of Fas in the nervous system was for a long period the object of relatively little study. It was known that Fas-L is quite widely expressed in the nervous system, but its presence was attributed to a role in conferring "immune privilege" on the CNS, through its ability to trigger the death of invading lymphocytes (French et al. 1996; French and Tschopp 1996; Saas et al. 1997). The situation has changed considerably over the last three years, and the role of Fas in both normal and pathological neuronal death has been addressed by several studies. We reviewed many of them recently elsewhere (Raoul et al. 1999b), and so have chosen here to put special emphasis on the model system in which our own work has been carried out: the developmental death of motoneurons in the spinal cord.

Roles for Fas in Motoneuron Death

PCD of Motoneurons During Development

Since the pioneering work of Hamburger and Levi-Montalcini, and thanks to a great extent to the important series of studies from the Oppenheim laboratory (see chapter by Oppenheim et al.), the naturally occurring death of approximately half of all spinal motoneurons during development has become a classical model system in which to study PCD of neurons. Motoneuron loss occurs soon after all the motoneurons have contacted their target muscle. Results of embryonic limb ablation experiments clearly show that all motoneurons require the presence of neurotrophic factors from the target muscle in order to survive. Moreover, the number of motoneurons lost during the cell death period can be reduced by administration of exogenous motoneuron survival factors. Based on such data, the neurotrophic hypothesis (Fig. 2 A) proposes that, at this critical stage of development, motoneurons compete for

Fig. 2. Models to explain naturally occurring loss of half of the motoneurons (MNs) initially produced. **A** The neurotrophic hypothesis: stochastic competition between motoneurons for access to limited supplies of muscle-derived neurotrophic factor. **B** Active killing of motoneurons by external triggers even in the presence of sufficient neurotrophic support

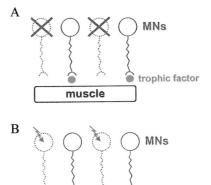

limited stocks of neurotrophic factor. Only those successful in gaining access to sufficient factor will subsequently survive.

Although the neurotrophic hypothesis may not be entirely sufficient to explain the regulation of motoneuron numbers during development (Pettmann and Henderson 1998), it has constituted an important intellectual framework for many different experiments and has led to the identification of a surprisingly large number of neurotrophic factors that can act on motoneurons (Henderson 1996; Oppenheim 1996). One aspect of the hypothesis that has been less discussed is the nature of the driving force for the death of those motoneurons that fail to gain enough trophic support. This aspect was the starting point of experiments on Fas in our laboratory (Raoul et al. 1999a, manuscript in preparation; Ugolini et al., manuscript in preparation).

Involvement of Fas/Fas Ligand in PCD of Motoneurons Deprived of Neurotrophic Factors in Culture

We tested the hypothesis that Fas activation might trigger death of motoneurons deprived of neurotrophic factors in vitro, a widely used model for competition for trophic support in vivo (Raoul et al. 1999a). We first showed that embryonic motoneurons co-express the death receptor Fas and its ligand Fas-L at the stage at which PCD is about to begin, both in vivo and in vitro. When cultured in the absence of trophic factors, many motoneurons die in culture within two days. Most (75%) of these were saved by soluble Fas-Fc, which blocks interactions between Fas and Fas-L, or by the caspase-8 inhibitor IETD.

To avoid problems with non-specific effects of the pharmacological agents used, we have also taken advantage of the availability of strains of mice that are deficient in Fas signalling: *lpr/lpr* mice, which express only low levels of Fas; *gld/gld* mice, which have a point mutation in FasL; and *Fas-/-* mice and mice over-expressing a dominant-negative form of the adaptor molecule FADD. When motoneurons purified from embryos of each strain were cultured in the absence of trophic support, we consistently observed that mutant motoneurons died >24 hr later than motoneurons from wildtype littermates. These results show that: 1) Fas signalling through Fas ligand is clearly involved in motoneuron PCD in these conditions, but 2) as in many other systems, if the Fas pathway is blocked, neurons will eventually activate other death pathways. This process makes evolutionary sense, since what has been selected for is the capacity to undergo PCD at the appropriate stage, rather than the use of any particular pathway.

Therefore, activation of Fas by endogenous Fas-L is one important driving force behind cell death induced by trophic deprivation. How does the Fas pathway become activated in the absence of neurotrophic factors, and how does signalling downstream of neurotrophic factor receptors prevent Fas-triggerd neuronal death? We observed that, as previously reported in other systems, removal of trophic factors led to strong up-regulation of *FasL*. The

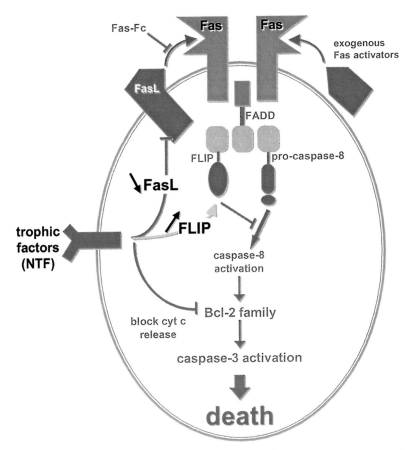

Fig. 3. Three different levels at which neurotrophic factors can act to prevent Fas-induced death of motoneurons: prevention of up-regulation of Fas ligand, up-regulation of the intracellular decoy FLIP, and signalling to the mitochondrion to prevent cytochrome *c* release

mechanisms for this are not known in motoneurons, but in other cell types the *FasL* gene has been shown to be regulated by the forkhead-related transcription factor FKHRL (Brunet et al. 1999), by c-Jun and by c-Myc (Faris et al. 1998; Brunner et al. 2000). Neurotrophic factors can block this stress-activated pathway in at least two ways that could each involve Akt kinase (Fig. 3). First, direct phosphorylation of BAD by Akt leads to inhibition of Bax function and therefore of cytochrome *c* release from mitochondria (Datta et al. 1997); this inhibition should slow PCD even if Fas is activated by FasL. Secondly, Akt can phosphorylate FKHRL, thereby preventing it from translocating to the nucleus to activate transcription of the *FasL* gene (Brunet et al. 1999).

Fas-Triggered PCD of Motoneurons in the Presence of Trophic Support

Having established a potential role of Fas in motoneurons following up-regulation of FasL in the absence of trophic support, we asked whether Fas signalling itself was completely blocked in the presence of trophic factors. If not, one could envisage a model whereby motoneuron PCD is triggered by death receptors without the need to invoke competition for trophic support (Fig. 2B). As an experimental model for this situation, we grew motoneurons in the presence of a cocktail of neurotrophic factors, and used exogenous Fas activators to mimic Fas activation by, say, neighbouring cells. Even in these highly favourable survival conditions, soluble FasL or anti-Fas antibodies triggered PCD of 40–50% of purified motoneurons over the following three to four days (Raoul et al. 1999 a). It is striking that the degree and time course of motoneuron loss observed in these culture conditions are very close to those of naturally occurring motoneuron PCD in the embryo (Yamamoto and Henderson 1999). Moreover, the results demonstrate that inhibition of cytochrome c release by neurotrophic factors is either not complete enough, or not in itself sufficient, to prevent Fas-triggered PCD, raising the possibility that other pathways may be involved.

Nevertheless, neurotrophic factors can overcome this type of Fas signalling as well: motoneurons cultured for three days (instead of one day) with neurotrophic factors became resistant to Fas activation, although they expressed normal levels of *Fas*. Correlatively, resistant motoneurons expressed high levels of FLIP, an endogenous inhibitor of caspase-8 activation (Raoul et al. 1999 a). This finding suggests a third pathway by which neurotrophic factors may chronically prevent Fas-triggered PCD (Fig. 3). If similar regulation occurs in vivo, then the levels of FLIP (rather than levels of Fas or FasL) may be critical in determining the time window during which motoneurons are lost during embryonic development; expression studies will be required to address this point. More generally, the existence of endogenous inhibitors of the Fas pathway may help to explain why relatively few accounts of neuron killing by Fas have been reported.

Signalling Pathways and Transcriptional Mechanisms Involved in Fas-Induced Neuronal Death

Since the pioneering work from the Johnson laboratory demonstrating that death of sympathetic neurons deprived of NGF could be blocked by translational and transcriptional inhibitors (Martin et al. 1988; Deckwerth and Johnson 1993), it has been clear that gene transcription is required for many types of neuronal PCD. This knowledge was for a period rather eclipsed by the many new and striking findings on post-translational signalling pathways. Nevertheless, it provides a rational explanation for why neuronal death often seems slow in comparison to that frequently observed in other systems. Thus, the fact that a trophically deprived neuron has first to up-regulate *FasL* before activating the cell death machinery per se could explain why typically this process takes 48 hr or more (Le-Niculescu et al. 1999). We were intrigued to observe that, even when Fas was activated by exogenous agents such as sFasL, a similar period was required before motoneuron loss could be observed (Raoul et al. 1999a). We have therefore decided to study potential transcriptional events downstream of Fas activation. The paradigm involving Fas activation in the presence of neurotrophic factors seems ideally suited to this.

What are some of the candidate genes that might be worth evaluating? One possible mechanism of stress-related motoneuron death that has been at the centre of considerable interest is the reaction of NO with superoxide anions to yield peroxynitrite (Estevez et al., 1998). This highly reactive species can then nitrosylate tyrosine residues in proteins and lead to cell death by poorly-understood mechanisms. The involvement of the peroxynitrite pathway in pathological motoneuron death in vivo is still controversial. However, we previously showed (Estevez et al., 1998) that motoneurons cultured in the absence of trophic support up-regulate the neuronal form of NOS (nNOS) within 18 hr. Moreover, the subsequent death of trophically-deprived neurons was prevented by incubation with either nNOS inhibitors, or with free radical scavengers such as MnTBAP. This suggested that the peroxynitrite pathway might be playing a role in motoneuron apoptosis in these conditions, the same paradigm in which we have demonstrated a role for Fas activation (Raoul et al., 1999a). This was further supported by the fact that motoneurons cultured without trophic support rapidly became immunoreactive for nitrotyrosine. One hypothesis worth investigation is therefore that Fas acts in motoneurons to transcriptionally up-regulate an element of this signalling pathway, the most likely candidate being the *nNOS* gene itself.

How might Fas activation be linked to transcriptional events downstream? In general, the signalling molecules best known for providing links between membrane receptors and transcriptional read-out are the stress-activated protein kinases, such as p38 or JNK. For instance, in some cell types and situations, the AP-1-associated factor ATF-2 (Mielke and Herdegen, 2000) can be activated following phosphorylation by the stress-activated protein kinase p38 (Ono and Han, 2000). It is tempting to speculate that the link from Fas to the stress-activated protein kinases occurs via the Daxx-ASK1 pathway discussed above (Fig. 1). In this context, it is striking that over-expression of constitutively active ASK1 in sympathetic neurons can trigger their death, and that trophic deprivation of the same cells leads to a 5-fold increase in ASK1 activity (Kanamoto et al., 2000).

The fact that Fas activation in neurons leads to a death process that is slow compared to other non-neuronal cell types argues for the existence of a transcriptionally-regulated pathway of the type hypothesized above. However, this does not rule out the probable use of the classical type 2 pathway of signalling through caspase-8 and BID (Fig. 1). Indeed, we have shown that the tetrapeptide IETD, which is a relatively specific inhibitor of caspase-8, completely inhibits the Fas-triggered death of motoneurons (Raoul et al., 1999a).

It is likely therefore that parallel pathways exist. These may be brought into play in different situations, or may simply ensure that the cell is capable of dying in response to the appropriate stimulus, even if one of the pathways is blocked.

Pathological Cell Death of Motoneurons in Patients with Amyotrophic Lateral Sclerosis

Amyotrophic lateral sclerosis (ALS) is a progressive neuro-degenerative disease in which motoneuron loss in the spinal cord and motor cortex leads to paralysis and death. Most cases of ALS are sporadic, but a small fraction are familial. One cause of familial ALS is a gain-of-function mutation in the *SOD1* gene, which triggers motoneuron death in a manner that is still not well understood. Mechanisms that have been proposed to explain sporadic ALS include: abnormal glutamate metabolism in spinal cord and motor cortex, environmental factors, and circulating cytotoxic molecules in serum and cerebrospinal fluid (reviewed in Henderson 1995; Brown 1997; Miller 1998).

Cytotoxicity of ALS sera has been reported by certain investigators, but not found by others. However, none of the published studies actually involved testing sera on motoneurons, the cells lost in human patients. We have demonstrated that circa 30% of sera from ALS patients contain macromolecules that rapidly and potently trigger the death of purified embryonic motoneurons from rat (Camu et al., in preparation). No comparable toxicity was found in any sera from normal or pathological controls, and sera toxic for motoneurons had little or no effect on survival of sensory neurons, which are spared in ALS patients. We were not able to identify the active macromolecules responsible for triggering motoneuron death. Recently, however, Yi et al. (2000) reported that 26% of ALS sera induced apoptosis of a human neuroblastoma cell line, and that most of these contained auto-antibodies to Fas. This finding raises the possibility of a wider involvement of Fas-triggered cell killing in human neurodegenerative disease.

Conclusions

There is a rapidly increasing body of evidence implicating Fas in both naturally occurring and pathological cell death in the nervous system. In the specific case of motoneuron cell death, it still remains to be demonstrated that Fas-triggered pathways are involved in vivo, and it will be important to investigate the effects of Fas inactivation on motoneuron death induced either by experimental lesion or genetic disease. Amidst the diversity of pathways involved in orchestrating Fas-dependent cell death, some appear to be neuron-specific and therefore in the future may constitute important avenues for new therapeutic approaches to neurodegenerative diseases such as ALS.

Acknowledgments. Work from the authors' laboratory was supported by IN-SERM, CNRS, Association Française contre les Myopathies (AFM), and ALS Association. Salary support was provided by the European Community (Marie Curie Programme) and the Ministère de la Recherche et de la Technologie.

References

Adachi M, Suematsu S, Kondo T, Ogasawara J, Tanaka T, Yoshida N, Nagata S (1995) Targeted mutation in the Fas gene causes hyperplasia in peripheral lymphoid organs and liver. Nat Genet 11:294–300

Ameisen JC (1996) The origin of programmed cell death. Science 272:1278–1279

Bredesen DE, Rabizadeh S (1997) p75^{NTR} and apoptosis: Trk-dependent and Trk-independent effects. Trends Neurosci 20:287–290

Brown RH, Jr (1997) Amyotrophic lateral sclerosis. Insights from genetics. Arch Neurol 54:1246–1250

Brunet A, Bonni A, Zigmond MJ, Lin MZ, Juo P, Hu LS, Anderson MJ, Arden KC, Blenis J, Greenberg ME (1999) Akt promotes cell survival by phosphorylating and inhibiting a Forkhead transcription factor. Cell 96:857–868

Brunner T, Mogil RJ, LaFace D, Yoo NJ, Mahboubi A, Echeverri F, Martin SJ, Force WR, Lynch DH, Ware CF, Green DR (1995) Cel-autonomous Fas (CD95)/Fas-ligand interaction mediates activation-induced apoptosis in T-cell hybridomas. Nature 373:441–444

Brunner T, Kasibhatla S, Pinkoski MJ, Frutschi C, Yoo NJ, Echeverri F, Mahboubi A, Green DR (2000) Expression of Fas ligand in activated T cells is regulated by c-Myc. J Biol Chem 275:9767–9772

Casaccia-Bonnefil P, Gu C, Chao MV (1999) Neurotrophins in cell survival/death decisions. Adv Exp Med Biol 468:275–282

Chang HY, Nishitoh H, Yang X, Ichijo H, Baltimore D (1998) Activation of apoptosis signal-regulating kinase 1 (ASK1) by the adapter protein Daxx. Science 281:1860–1863

Chinnaiyan AM, O'Rourke K, Tewari M, Dixit VM (1995) FADD, a novel death domain-containing protein, interacts with the death domain of Fas and initiates apoptosis. Cell 81:505–512

Datta SR, Dudek H, Tao X, Masters S, Fu H, Gotoh Y, Greenberg ME (1997) Akt phosphorylation of BAD couples survival signals to the cell-intrinsic death machinery. Cell 91:231–241

Deckwerth TL, Johnson EM, Jr. (1993) Temporal analysis of events associated with programmed cell death (apoptosis) of sympathetic neurons deprived of nerve growth factor. J Cell Biol 123:1207–1222

Deckwerth TL, Elliott JL, Knudson CM, Johnson EM, Jr, Snider WD, Korsmeyer SJ (1996) BAX is required for neuronal death after trophic factor deprivation and during development. Neuron 17:401–411

Desagher S, Osen-Sand A, Nichols A, Eskes R, Montessuit S, Lauper S, Maundrell K, Antonsson B, Martinou JC (1999) Bid-induced conformational change of Bax is responsible for mitochondrial cytochrome *c* release during apoptosis. J Cell Biol 144:891–901

Dhein J, Walczak H, Baumler C, Debatin KM, Krammer PH (1995) Autocrine T-cell suicide mediated by APO-1/(Fas/CD95). Nature 373:438–441

Di Cristofano A, Pandolfi PP (2000) The multiple roles of PTEN in tumor suppression. Cell 100:387–390

Di Cristofano A, Kotsi P, Peng YF, Cordon-Cardo C, Elkon KB, Pandolfi PP (1999) Impaired Fas response and autoimmunity in Pten +/- mice. Science 285:2122–2125

Eskes R, Desagher S, Antonsson B, Martinou JC (2000) Bid induces the oligomerization and insertion of Bax into the outer mitochondrial membrane. Mol Cell Biol 20:929–935

Estevez AG, Spear N, Manuel SM, Radi R, Henderson CE, Barbeito L, Beckman JS (1998) Nitric oxide and superoxide contribute to motor neuron apoptosis induced by trophic factor deprivation. J Neurosci 18:923–931

Faris M, Kokot N, Latinis K, Kasibhatla S, Green DR, Koretzky GA, Nel A (1998) The c-Jun N-terminal kinase cascade plays a role in stress-induced apoptosis in Jurkat cells by up-regulating Fas ligand expression. J Immunol 160:134–144

Frade JM, Barde YA (1998) Microglia-derived nerve growth factor causes cell death in the developing retina. Neuron 20:35–41

Frade JM, Barde YA (1999) Genetic evidence for cell death mediated by nerve growth factor and the neurotrophin receptor p75 in the developing mouse retina and spinal cord. Development 126:683–690

Frade JM, Rodriguez-Tebar A, Barde YA (1996) Induction of cell death by endogenous nerve growth factor through its p75 receptor. Nature 383:166–168

French LE, Tschopp J (1996) Constitutive Fas ligand expression in several non-lymphoid mouse tissues: implications for immune-protection and cell turnover. Behring Inst Mitt, pp 156–160

French LE, Hahne M, Viard I, Radlgruber G, Zanone R, Becker K, Muller C, Tschopp J (1996) Fas and Fas ligand in embryos and adult mice: ligand expression in several immune-privileged tissues and coexpression in adult tissues characterized by apoptotic cell turnover. J Cell Biol 133:335–343

Golstein P (2000) Signal transduction. FasL binds preassembled Fas. Science 288:2328–2329

Henderson CE (1995) Neurotrophic factors as therapeutic agents in amyotrophic lateral sclerosis: potential and pitfalls. Adv Neurol 68:235–240

Henderson CE (1996) Role of neurotrophic factors in neuronal development. Curr Opin Neurobiol 6:64–70

Jacobson MD, Weil M, Raff MC (1997) Programmed cell death in animal development. Cell 88:347–354

Kanamoto T, Mota M, Takeda K, Rubin LL, Miyazono K, Ichijo H, Bazenet CE (2000) Role of apoptosis signal-regulating kinase in regulation of the c-Jun N-terminal kinase pathway and apoptosis in sympathetic neurons. Mol Cell Biol 20:196–204

Kaplan DR, Miller FD (1997) Signal transduction by the neurotrophin receptors. Curr Opin Cell Biol 9:213–221

Knudson CM, Tung KS, Tourtellotte WG, Brown GA, Korsmeyer SJ (1995) Bax-deficient mice with lymphoid hyperplasia and male germ cell death. Science 270:96–99

Le-Niculescu H, Bonfoco E, Kasuya Y, Claret FX, Green DR, Karin M (1999) Withdrawal of survival factors results in activation of the JNK pathway in neuronal cells leading to Fas ligand induction and cell death. Mol Cell Biol 19:751–763

Li H, Zhu H, Xu CJ, Yuan J (1998) Cleavage of BID by caspase 8 mediates the mitochondrial damage in the Fas pathway of apoptosis. Cell 94:491–501

Luo X, Budihardjo I, Zou H, Slaughter C, Wang X (1998) Bid, a Bcl2 interacting protein, mediates cytochrome c release from mitochondria in response to activation of cell surface death receptors. Cell 94:481–490

Martin DP, Schmidt RE, DiStefano PS, Lowry OH, Carter JG, Johnson EM, Jr (1988) Inhibitors of protein synthesis and RNA synthesis prevent neuronal death caused by nerve growth factor deprivation. J Cell Biol 106:829–844

Martinou JC, Dubois-Dauphin M, Staple JK, Rodriguez I, Frankowsky H, Missotten M, Albertini P, Talabot D, Catsicas S, Pietra C, Huarte J (1994) Overexpression of bcl-2 in transgenic mice protects neurons from naturally occurring cell death and experimental ischaemia. Neuron 13:1017–1030

Mielke K, Herdegen T (2000) JNK and p38 stress kinases – degenerative effectors of signal-transduction-cascades in the nervous system. Prog Neurobiol 61:45–60

Miller RG (1998) New approaches to therapy of amyotrophic lateral sclerosis. West J Med 168:262–263

Muzio M, Chinnaiyan AM, Kischkel FC, O'Rourke K, Shevchenko A, Ni J, Scaffidi C, Bretz JD, Zhang M, Gentz R, Mann M, Krammer PH, Peter ME, Dixit VM (1996) FLICE, a novel FADD-homologous ICE/CED-3-like protease, is recruited to the CD95 (Fas/APO-1) death-inducing signaling complex. Cell 85:817–827

Nagata S (1997) Apoptosis by death factor. Cell 88:355–365

Nagata S, Golstein P (1995) The Fas death factor. Science 267:1449–1456

Nishitoh H, Saitoh M, Mochida Y, Takeda K, Nakano H, Rothe M, Miyazono K, Ichijo H (1998) ASK1 is essential for JNK/SAPK activation by TRAF2. Mol Cell 2:389–395

Ono K, Han J (2000) The p38 signal transduction pathway: activation and function. Cell Signal 12:1–13

Oppenheim RW (1991) Cell death during development of the nervous system. Annu Rev Neurosci 14:453–501

Oppenheim RW (1996) Neurotrophic survival molecules for motoneurons: an embarrassment of riches. Neuron 17:195–197

Pettmann B, Henderson CE (1998) Neuronal cell death. Neuron 20:633–647

Raoul C, Henderson CE, Pettmann B (1999a) Programmed cell death of motoneurons triggered through the Fas death receptor. J Cell Biol 147:1049–1061

Raoul C, Pettmann B, Henderson CE (1999b) Active killing of neurons during development and following stress: a role for p75NTR and Fas? Curr Opin Neurobiol 10:111–117

Rondi-Reig L, Lohof A, Dubreuil YL, Delhaye-Bouchaud N, Martinou JC, Caston C, Mariani J (1999) Hu-Bcl-2 transgenic mice with supernumerary neurons exhibit timing impairment in a complex motor task. Eur J Neurosci 11:2285–2290

Saas P, Walker PR, Hahne M, Quiquerez AL, Schnuriger V, Perrin G, French L, Van Meir EG, de Tribolet N, Tschopp J, Dietrich PY (1997) Fas ligand expression by astrocytoma in vivo: maintaining immune privilege in the brain? J Clin Invest 99:1173–1178

Stambolic V, Suzuki A, de la Pompa JL, Brothers GM, Mirtsos C, Sasaki T, Ruland J, Penninger JM, Siderovski DP, Mak TW (1998) Negative regulation of PKB/Akt-dependent cell survival by the tumor suppressor PTEN. Cell 95:29–39

Sun H, Lesche R, Li DM, Liliental J, Zhang H, Gao J, Gavrilova N, Mueller B, Liu X, Wu H (1999) PTEN modulates cell cycle progression and cell survival by regulating phosphatidylinositol 3,4,5-triphosphate and Akt/protein kinase B signaling pathway. Proc Natl Acad Sci USA 96:6199–6204

Takahashi T, Tanaka M, Brannan CI, Jenkins NA, Copeland NG, Suda T, Nagata S (1994) Generalized lymphoproliferative disease in mice, caused by a point mutation in the Fas ligand. Cell 76:969–976

Watanabe-Fukunaga R, Brannan CI, Copeland NG, Jenkins NA, Nagata S (1992) Lymphoproliferation disorder in mice explained by defects in Fas antigen that mediates apoptosis. Nature 356:314–317

Yaginuma H, Tomita M, Takashita N, McKay SE, Cardwell C, Yin QW, Oppenheim RW (1996) A novel type of programmed neuronal death in the cervical spinal cord of the chick embryo. J Neurosci 16:3685–3703

Yamamoto Y, Henderson CE (1999) Patterns of programmed cell death in populations of developing spinal motoneurons in chicken, mouse, and rat. Dev Biol 214:60–71

Yi FH, Lautrette C, Vermot-Desroches C, Bordessoule D, Couratier P, Wijdenes J, Preud'homme JL, Jauberteau MO (2000) In vitro induction of neuronal apoptosis by anti-Fas antibody-containing sera from amyotrophic lateral sclerosis patients. J Neuroimmunol 109:211–220

Events that Commit Neurons
to Die After Trophic Factor Deprivation

G. V. Putcha, M. Deshmukh, and E. M. Johnson, Jr.

Summary

Nerve growth factor (NGF) deprivation in neonatal sympathetic neurons induces two parallel processes: 1) a protein synthesis-dependent, caspase-independent translocation of BAX from the cytosol to mitochondria, followed by mitochondrial membrane integration and loss of cytochrome c; and 2) the development of competence-to-die, which requires neither macromolecular synthesis nor BAX expression. Activation of both signaling pathways is required for caspase activation and apoptosis in immature sympathetic neurons.

The identities of the gene products required for the translocation of BAX and for its subsequent integration and mediation of cytochrome c release (presumably via multimerization and pore formation) remain unknown. Recently, we have identified one such "thanatin": BIM, a member of the BH3-only, proapoptotic subfamily of the BCL-2 family of proteins.

NGF deprivation rapidly induces expression of the extra-long splice variant of BIM, BIM_{EL}, upstream of the BAX/BCL-2 and caspase checkpoints. Other findings indicate that the induction of BIM may constitute a unique hallmark of neuronal apoptosis. Moreover, *Bim* deletion confers transient protection against cytochrome c release and neuronal apoptosis, suggesting that BIM, and perhaps other BH3-only proteins, serve partially redundant functions upstream of BAX-mediated cytochrome c release and caspase activation.

The events responsible for committing sympathetic neurons to die after trophic factor withdrawal remain unclear. NGF-deprived cells become committed to die, as measured by the inability of cells to be rescued by NGF readdition, at the time of cytochrome c release. However, caspase inhibition by either pharmacological or genetic means extends commitment to death from the point of cytochrome c release to the subsequent point of mitochondrial depolarization.

Introduction

Programmed cell death (PCD) is an evolutionarily conserved and genetically regulated process that is critical to the development and maintenance of many tissues. Deficiencies in PCD underlie some forms of oncogenesis,

Henderson/Green/Mariani/Christen (Eds.)
Neuronal Death by Accident or by Design
© Springer-Verlag Berlin Heidelberg 2001

whereas excessive cell death may contribute to several pathological conditions, including stroke, autoimmune disorders, and certain neurodegenerative diseases (Thompson 1995). Cells undergoing PCD exhibit morphological and biochemical changes characteristic of apoptosis, including cytoplasmic shrinkage, plasma-membrane blebbing, chromatin condensation, and DNA fragmentation (Kerr et al. 1972). Non-professional phagocytes eventually engulf the dying cells, thereby preventing an inflammatory response.

Genetic and biochemical studies have identified several critical regulators of PCD in mammals: caspases, Apaf-1, and the BCL-2 family of proteins. These proteins are the mammalian homologs of the *ced-3*, *ced-4*, and *ced-9* gene products, respectively, which are required for the proper regulation of PCD in *Caenorhabditis elegans* (Ellis et al. 1991; Metzstein et al. 1998).

The BCL-2 family of proteins consists of both proapoptotic (e.g., BAX and BAK) and antiapoptotic (e.g., BCL-2 and BCL-x_L) members. Family members exhibit sequence homology in at least one of four BCL-2 homology (BH) domains, which can mediate protein-protein interactions (Oltvai and Korsmeyer 1994), including the formation of certain homodimers and heterodimers. The ratio of proapoptotic to antiapoptotic BCL-2 family members within a particular cell may determine whether the cell will die in response to a death signal (Oltvai et al. 1993); however, recent evidence also suggests that family members can function autonomously to regulate cell death (Knudson and Korsmeyer 1997). Although the precise mechanisms by which BCL-2 family proteins regulate survival remain enigmatic, overexpression of antiapoptotic family members inhibits – and overexpression of proapoptotic members promotes – the loss of mitochondrial cytochrome *c* during cell death (Kluck et al. 1997; Yang et al. 1997; Jürgensmeier et al. 1998). Moreover, the three-dimensional structure of BCL-x_L shows similarities to the pore-forming domains of bacterial toxins such as diphtheria toxin and the colicins (Muchmore et al. 1996). Accordingly, BCL-2, BCL-x_L, and BAX can form ion channels in synthetic lipid membranes (Antonsson et al. 1997; Minn et al. 1997; Schendel et al. 1997; Schlesinger et al. 1997). These channels are characterized by multiconductance states, pH sensitivity, voltage gating, and poor ion selectivity. Whether BCL-2 family members form pores in mitochondria in intact cells, and what regulates this activity, are unclear; however, these findings suggest that these proteins may either form or regulate the formation of channels in mitochondrial membranes.

Recently discovered within the BCL-2 family is a new subfamily of proapoptotic death "ligands" that share homology only within the BH3 domain. The physiological functions and regulation of this BH3-only subfamily, which includes mammalian BAD, BID, BIK/NBK, BIM, BLK, HRK/DP5 and nematode EGL-1, are gradually becoming illuminated. These proteins can regulate cell death both autonomously and nonautonomously and may confer cell-type and stimulus specificity to various apoptotic insults. For example, not only can BH3-only proteins such as BID and BIM regulate cell death indirectly through modulation of the activities of the BAX and BCL-2 subfamilies (Desagher et al. 1999; Puthalakath et al. 1999; Eskes et al. 2000), but at least one of these proteins (i.e., BID in Fas-mediated death) can also directly activate the apoptotic effector machinery (Li et al. 1998; Luo et al. 1998; Yin et al. 1999).

In mammals, signaling cascades culminating in cell death can be divided into two broad categories: the "intrinsic" (i.e., apoptosome) and the "extrinsic" (i.e., death receptor and perforin/granzyme) pathways (Stennicke and Salvesen 2000). Current evidence suggests the following model for activation of the "intrinsic" pathway of apoptosis. A death signal induces integration of a proapoptotic BCL-2 family member in the mitochondrial outer membrane, followed by the release of mitochondrial proteins, such as cytochrome c (Liu et al. 1996), through an unknown mechanism that may involve a channel formed by the proapoptotic BCL-2 family proteins and/or the permeability transition pore (PTP). Once released, cytochrome c forms a complex with Apaf-1 and procaspase-9, which becomes activated in the presence of ATP or dATP, resulting in further caspase activation, cleavage of cellular substrates, and cell death (Liu et al. 1996; Li et al. 1997; Zou et al. 1997).

In the nervous system, PCD occurs during both developmental and pathological processes. During development, 20–80% of all neurons produced during embryogenesis die before reaching adulthood (Oppenheim 1991), ostensibly to match the number of innervating neurons with the size of the target tissue. The neurotrophic hypothesis proposes that neurons compete with each other for limited amounts of survival-promoting neurotrophins, thereby ensuring appropriate target innervation. Accordingly, significant neuronal PCD occurs during the period of target innervation.

One of the most extensively studied models of neuronal PCD is nerve growth factor (NGF) deprivation in neonatal sympathetic neurons. From approximately embryonic day 16 (E16) to postnatal day 7 (P7), these neurons require NGF for survival (Coughlin and Collins 1985). Administration of NGF antiserum during this period induces extensive loss of sympathetic neurons (Levi-Montalcini and Booker 1960). This cell death can be recapitulated in vitro. Neonatal sympathetic neurons maintained in culture for four to six days in the presence of NGF undergo an apoptotic cell death within 48 hours of trophic factor withdrawal (Edwards et al. 1991; Deckwerth and Johnson 1993). This death, which requires de novo protein synthesis (Martin et al. 1988), *Bax* expression (Deckwerth et al. 1996), caspase activation (Deshmukh et al. 1996; Troy et al. 1996; McCarthy et al. 1997), and the development of competence-to-die (Deshmukh and Johnson 1998), is blocked by neuroprotective agents, such as KCl and cAMP (Rydel and Greene 1988; Koike et al. 1989; Edwards et al. 1991; Deckwerth and Johnson 1993).

BAX Translocation is a Critical Event in Neuronal Apoptosis

Despite widespread expression of BAX in vivo, *Bax*-deficient mice exhibit relatively modest defects in nonneuronal lineages, perhaps because of functional redundancy with other proapoptotic BCL-2 family members (Knudson et al. 1995). In contrast, their phenotype in neuronal lineages, whether central or peripheral, is dramatic (Deckwerth et al. 1996; Miller et al. 1997; Shindler et al. 1997; White et al. 1998). For example, despite coexpression of multiple proapoptotic BCL-2 family members, cell death induced by trophic factor deprivation in sympathetic and motor neurons is remarkably dependent on BAX alone.

Recent reports indicate that the regulation of the proapoptotic activity of BAX involves its subcellular localization. In several nonneuronal models of cell death, overexpression of *Bax* and/or certain death signals induce translocation of BAX from the cytosol to mitochondria (Hsu et al. 1997; Wolter et al. 1997), followed by cytochrome *c* release, mitochondrial dysfunction, and cell death (Gross et al. 1998; Rosse et al. 1998; Saikumar et al. 1998; Finucane et al. 1999). However, these nonneuronal models of PCD may not require either endogenous BAX expression (Knudson et al. 1995) or caspase activation (Xiang et al. 1996) for cell death. Moreover, *Bax* overexpression per se drives its mitochondrial translocation (Rosse et al. 1998) and induces a caspase-independent, nonapoptotic cell death (Xiang et al. 1996).

Neonatal sympathetic neurons deprived of NGF undergo an apoptotic cell death that requires macromolecular synthesis (Martin et al. 1988), caspase activation (Deshmukh et al. 1996; Troy et al. 1996; McCarthy et al. 1997), and the development of competence-to-die (Deshmukh and Johnson 1998). Most important, unlike all known nonneuronal models of cell death, sympathetic neurons absolutely require the endogeneous expression of BAX to undergo apoptosis induced by trophic factor withdrawal (Deckwerth et al. 1996). However, because neither the mRNA nor protein level of BAX increases during NGF deprivation-induced apoptosis (Greenlund et al. 1995), BAX must be regulated posttranslationally.

We found that NGF deprivation in sympathetic neurons induces the subcellular redistribution of BAX from the cytosol to mitochondria (Putcha et al. 1999). Moreover, contrary to some reports in nonneuronal models of cell death (Goping et al. 1998; Desagher et al. 1999), this redistribution reflects the actual translocation of BAX from the cytosol to mitochondria, where BAX then integrates into the mitochondrial outer membrane (Antonsson et al. 2000; Eskes et al. 2000; Putcha et al. 2000). BAX translocation is followed rapidly, but not instantaneously, by cytochrome *c* release and commitment-to-die (Putcha et al. 1999). Taken together with our previous observations (Putcha et al. 2000), these findings indicate that BAX translocation per se is necessary for cytochrome *c* release, caspase activation, and apoptosis induced by trophic factor deprivation in sympathetic neurons.

The precise mechanisms by which BAX mediates loss of mitochondrial cytochrome *c* remain unclear. Recent evidence suggests that two potentially interrelated mechanisms may mediate this release: the calcium-inducible, cyclosporin A (CsA)-sensitive PTP and a Mg^{2+}-sensitive, CsA-insensitive pathway that depends on BAX (or perhaps more generally on proapoptotic BCL-2 family members; Eskes et al. 1998; Marzo et al. 1998; Narita et al. 1998). If BAX and the PTP mediate two biochemically and molecularly distinct pathways for cytochrome *c* release, four observations argue against a significant role for the PTP in sympathetic neuronal death. First, primary sympathetic neurons absolutely require BAX alone for cytochrome c release (Deshmukh and Johnson 1998; Putcha et al. 2000), caspase activation, and apoptosis (Deckwerth et al. 1996). Second, mitochondria in NGF-deprived neurons do not exhibit the ultrastructural changes, such as swelling and outer membrane rupture, that should accompany permeability transition (Martin et al. 1988; Martinou et al. 1999). Third, non-toxic concentrations of cyclosporin A neither prevent nor delay cytochrome c release or cell death induced by NGF deprivation in murine sympathetic neurons (unpublished observations). Fourth, we find that a significant delay occurs between cytochrome c release and loss of mitochondrial membrane potential (see below; Desmukh and Johnson 2000). Therefore, our data suggest that permeability transition does not mediate cytochrome c release during trophic factor deprivation-induced cell death in neurons.

Finally, although neither the cytosol-to-mitochondria redistribution of BAX (Putcha et al. 1999, 2000) nor the reciprocal translocation of cytochrome c requires caspase activity (Deshmukh and Johnson 1998, 2000; Neame et al. 1998; Martinou et al. 1999; Putcha et al. 1999, 2000), both processes require macromolecular synthesis (Deshmukh and Johnson 1998; Putcha et al. 1999). Thus, trophic factor withdrawal in neonatal sympathetic neurons requires a protein synthesis-dependent, caspase-independent translocation of BAX from the cytosol to mitochondria, followed by mitochondrial membrane integration and loss of cytochrome c (Putcha et al. 1999, 2000). However, the identities of the gene products required for BAX-mediated cytochrome c release are unknown.

Induction of BIM, a Proapoptotic BH3-only BCL-2 Family Member, Is Critical for Neuronal Apoptosis

Since the initial report demonstrating that NGF deprivation-induced apoptosis in sympathetic neurons requires macromolecular synthesis (Martin et al. 1988), the search for those genes, dubbed "thanatins" (Johnson et al. 1989), that are upregulated during the death process and critical for its execution has been, and continues to be, an area of active investigation. Until recently, genes reported to be induced during neuronal death fell primarily into three distinct categories: transcription factors of the AP-1 subfamily (e.g., *c-jun* and *c-fos*), cell cycle regulators (e.g., *c-myb* and *cyclinD1*), and extracellular matrix proteases (e.g., *transin* and *collagenase*; Estus et al. 1994; Freeman et al. 1994). However, no component of the core apoptotic machinery is induced by trophic factor withdrawal in these neurons.

More recently, Imaizumi et al. (1997) and Inohara et al. (1997) reported the cloning of HRK/DP5, a proapoptotic member of the BCL-2 protein family that is induced by NGF deprivation in sympathetic neurons. Overexpression of HRK/DP5, which contains only BH3 and transmembrane domains, causes cell death in both neuronal and nonneuronal cells that can be attenuated by co-overexpression of BCL-2 or BCL-x$_L$, possibly through direct BH3 domain-mediated interactions. Furthermore, although the subcellular distribution of endogenous HRK/DP5 is unknown, overexpression of epitope-tagged HRK/DP5 indicates localization to intracytoplasmic membranes. Thus, HRK/DP5 may be a bona fide thanatin; however, the physiological function of endogenous HRK/DP5 remains unclear. We have identified and functionally characterized another molecule that satisfies the criteria for a thanatin: BIM, which like HRK/DP5 is a member of the emerging BH3-only proapoptotic subfamily of the BCL-2 family of proteins.

Initially cloned as a BCL-2 interacting mediator of cell death from a bacteriophage λ cDNA library derived from the p53$^{-/-}$ T lymphoma cell line KO52DA20, BIM is expressed in many cell types as three alternatively spliced isoforms: bim_S, bim_L, and bim_{EL} (O'Connor et al. 1998). Analysis of protein expression indicates that BIM$_L$ and BIM$_{EL}$ are the predominant isoforms expressed in cell lines and normal tissues (O'Reilly et al. 1998, 2000). Although BIM bears little sequence homology to any other proteins in current databases, sequence analysis reveals a stretch of nine amino acids corresponding to a BH3 domain and a C-terminal hydrophobic region presumably serving as a transmembrane anchor. Accordingly, subcellular fractionation and immunocytochemical analysis of cell lines stably or transiently overexpressing epitope-tagged BIM$_L$ indicates localization to intracytoplasmic membranes. Furthermore, overexpression of individual isoforms demonstrates that all three splice variants are potent inducers of cell death that can be antagonized by co-overexpression of antiapoptotic BCL-2 family members or caspase inhibitors (O'Connor et al. 1998). The level of cytotoxicity correlates inversely with isoform length (i.e., BIM$_S$ > BIM$_L$ > BIM$_{EL}$), suggesting that the additional regions present in BIM$_L$ and BIM$_{EL}$ may attenuate proapoptotic activity. In FDC-P1 cells (an IL-3-dependent promyelomonocytic cell line) stably overexpressing both BIM$_L$ and BCL-2, certain death signals, such as IL-3 withdrawal, UV irradiation, or staurosporine, induce the translocation of a complex consisting of BIM$_L$ and the LC8 cytoplasmic-dynein light chain from the microtubule-associated dynein motor complex to the mitochondria. There, BIM interacts with BCL-2, thereby inhibiting the latter's antiapoptotic activity and ensuring the ultimate demise of the cell (Puthalakath et al. 1999).

Targeted deletion of *Bim* by homologous recombination induces embryonic lethality by E9.5 in ~65% of *Bim*$^{-/-}$ mice with variable penetrance strongly affected by genetic background (Bouillet et al. 1999; O'Reilly et al. 2000). Moreover, *Bim* deficiency perturbs leukocyte homeostasis, as indicated by changes in B and T lymphocytes, granulocytes, and monocytes. Analysis of lymphocytes from *Bim*-deficient mice reveals selective apoptotic deficits. For example, pre-T (CD4$^+$CD8$^+$), pre-B (B220$^+$sIg$^-$CD43$^-$), and activated T lymphoblasts exhibit decreased sensitivity to certain death stimuli (e.g., IL-2 withdrawal, dexamethasone, and γ-irradiation) but not to others (e.g., FasL, etoposide, and phorbol ester), suggesting that different BH3-only proteins may contribute to apoptosis in a cell-type and stimulus-specific manner. Finally, 55% of live-born $Bim^{-/-}$ and 35% of $Bim^{+/-}$ animals have shortened life spans of approximately one year because of progressive lymphadenopathy with prominent plasmacytosis and systemic autoimmune disease.

Neonatal sympathetic neurons, like all neuronal populations examined (including cerebellar granule and dorsal root ganglion neurons) express mRNA for all three splice variants of BIM, yet express primarily BIM_{EL} protein (see also O'Reilly et al. 2000). Like HRK/DP5 (Imaizumi et al. 1997; Inohara et al. 1997), BIM normally appears to be expressed at relatively low levels in neurons both in vivo and in vitro. However, apoptotic insults significantly induce BIM_{EL} expression. For example, disparate death signals such as NGF withdrawal, potassium deprivation, and sciatic nerve axotomy induce BIM_{EL} in sympathetic, cerebellar granule, and dorsal root ganglion neurons, respectively. By contrast, death-inducing stimuli such as dexamethasone, staurosporine, and γ-irradiation do not induce expression of any BIM isoform in nonneuronal cells (G. V. Putcha and E. M. Johnson, unpublished observations; O'Connor et al. 1998). Taken together, these observations suggest that induction of BIM_{EL}, and perhaps of HRK/DP5 (Imaizumi et al. 1997), constitutes a hallmark of neuronal, but not nonneuronal, apoptosis.

Analysis of neurons from Bim-deficient mice underscores the importance of BIM function in neuronal apoptosis. In both model systems examined – NGF deprivation in sympathetic neurons and potassium deprivation in cerebellar granule neurons – Bim deletion confers protection against the loss of mitochondrial cytochrome c and neuronal apoptosis. However, the protection conferred by Bim deficiency is transient: $Bim^{-/-}$ sympathetic and cerebellar granule neurons ultimately release cytochrome c and undergo an apoptotic cell death. In contrast, targeted deletion of another BH3-only subfamily member, BAD, neither prevents nor alters the kinetics of commitment to cell death in both models of neuronal apoptosis (G. V. Putcha, K. L. Moulder, S. J. Korsmeyer, and E. M. Johnson, Jr., unpublished observations).

Furthermore, BIM_{EL} induction occurs in Bcl-2-overexpressing, Bax-deficient, and caspase inhibitor-treated sympathetic neurons with kinetics essentially identical to those in wild-type neurons, indicating that this event lies upstream of (or in parallel with) the BAX/BCL-2 and caspase checkpoints. Moreover, subcellular fractionation and extraction studies indicate that, when induced, endogenous BIM_{EL} exists as an integral membrane protein in mitochondria, suggesting that BIM_{EL} – and perhaps HRK/DP5 – probably does not mediate the translocation of BAX to mitochondria, but instead contributes to the subsequent integration of BAX into the mitochondrial outer membrane and/or BAX-mediated cytochrome c release. Although we are unclear exactly how BIM and presumably HRK/DP5 execute these functions, we propose two possibilities that need not be mutually exclusive. First, BIM and/or HRK/DP5 may directly interact with and inactivate antiapoptotic BCL-2 family members such as BCL-2 and BCL-x_L at the mitochondrial outer membrane (OM), thereby freeing BAX to multimerize and integrate into the OM, forming pores through which intermembrane space proteins such as cytochrome c may pass. [Recent evidence suggests that the multimerization of

BAX may either contribute to or be required for OM insertion and pore formation (Antonsson et al. 2000; Saito et al. 2000).] Alternatively, BH3-only proteins such as BIM and HRK/DP5 may themselves constitute members of a multiprotein pore complex in which BAX is an obligate component and which is responsible for the release of intermembrane space proteins such as cytochrome c.

In sum, our observations suggest that BIM – and perhaps HRK/DP5 – serves at least partially redundant functions upstream of the BAX/BCL-2 and caspase checkpoints, ultimately contributing to BAX-mediated cytochrome c release, caspase activation, and cell death. Based on our observations, we predict that neurons from *Hrk/Dp5*-deficient mice, if viable, will exhibit a transiently protective phenotype like that seen with $Bim^{-/-}$ cells. Cytochrome c release and caspase activation will be delayed, but not prevented, as in $Bim^{-/-}$ neurons. Moreover, reproduction of the neuronal phenotype seen in $Bax^{-/-}$ mice, in which cytochrome c release and caspase activation are completely prevented (Deshmukh and Johnson 1998; Putcha et al. 2000), may require inactivation of at least both BIM and HRK/DP5.

Caspase Inhibition Extends the Commitment to Neuronal Death Beyond Cytochrome c Release to the Point of Mitochondrial Depolarization

Recent reports suggest that cytochrome c release represents the "point of no return", when a cell undergoing apoptosis via the intrinsic pathway becomes irreversibly committed to die. Such loss of cytochrome c leads to mitochondrial depolarization and eventually compromises cellular respiration, culminating in cell death. As indicated above, the precise mechanism by which cytochrome c is released from mitochondria is unknown; however, the two most prominent competing hypotheses are 1) selective OM permeabilization because of pore formation by proapoptotic BCL-2 family members, such as BAX; and 2) nonselective OM rupture secondary to mitochondrial permeability transition (MPT) or closure of the voltage-dependent anion channel (VDAC; Kroemer and Reed 2000). Either mechanism eventually leads to cytochrome c release and mitochondrial depolarization; however, the MPT-dependent mechanism predicts that loss of mitochondrial transmembrane potential precedes cytochrome c release whereas the converse is predicted by the selective OM permeabilization hypothesis. Although some examples of mitochondrial depolarization preceding, or occurring concurrently with, cytochrome c release have been reported (Bradham et al. 1998; Heiskanen et al. 1999), other reports indicate that mitochondrial depolarization *follows* cytochrome c release (McCarthy et al. 1997; Ohta et al. 1997; Amarante-Mendes et al. 1998a; Brunet et al. 1998; Gibson 1999). In NGF-deprived mouse sympathetic neurons, we found that loss of cytochrome c precedes mitochondrial depolarization by ~30 hours. (Note that these experiments were done in the presence of a caspase inhibitor, since in the absence of such inhibition, mitochondrial depolarization, caspase activation, and commitment to death follow cytochrome c release very rapidly.)

Finally, the events that determine the point at which a cell becomes committed to die in the presence of caspase inhibitors are unknown. Since mitochondrial events such as cytochrome c release and inner membrane depolarization precede caspase activation in most models of apoptosis, cells saved by caspase inhibitors may eventually die because of the deterioration of critical cellular functions, particularly oxidative phosphorylation. Indeed, in *all* reported cases in which the initial death-inducing stimulus remains unabated, cells die even in the presence of caspase inhibitors via a slower, apparently nonapoptotic, cell death pathway. However, when the initiating insult is aborted and a normal physiological milieu restored, a fundamental difference between mitotic and postmitotic cells is uncovered. In the former, caspase inhibition blocks the characteristic features of apoptosis but not the timing of commitment to death (McCarthy et al. 1997; Miller et al. 1997; Ohta et al. 1997; Amarante-Mendes et al. 1998b; Brunet et al. 1998; Green and Reed 1998; Gibson 1999). In contrast, in postmitotic cells such as neurons, caspase inhibitors delay the commitment to death for a finite, but real, period of time, permitting a therapeutic "window of opportunity" during which these cells can be rescued. For example, when sympathetic neurons are deprived of NGF in the presence of caspase inhibitors for some time and then NGF is added back to the cultures, these cells can resume somal growth (Deshmukh et al. 1996), replenish their mitochondria with cytochrome c (Martinou et al. 1999), and recover electrophysiological function (Werth et al. 2000). However, the precise duration of this therapeutic window – and the molecular and biochemical determinants of when it closes – are unknown. Clearly, answers to such questions are crucial to understanding the clinical usefulness of caspase inhibition.

When we examined these questions in NGF-deprived mouse sympathetic neurons, we found that caspase inhibition (by both pharmacological and genetic means) extends the timing of commitment to death in NGF-deprived mouse sympathetic neurons by ~25–30 hours, from the point of cytochrome c release to the subsequent point of mitochondrial depolarization (Deshmukh and Johnson 2000). Specifically, in either caspase inhibitor-saved or *caspase-9*-deficient sympathetic neurons deprived of NGF, cytochrome c is lost but cells do not become committed to die until they also lose mitochondrial membrane potential. In fact, the maintenance of mitochondrial membrane potential, as assessed by staining with the potentiometric dye Mitotracker Orange (Molecular Probes, Eugene, OR), predicts those neurons that can be rescued by NGF readdition.

In summary, in the absence of caspase inhibitors, NGF-deprived sympathetic neurons become committed to die when cytochrome c is released from mitochondria into the cytosol, where cytochrome c facilitates apoptosome formation and caspase activation, culminating in an apoptotic cell death. In contrast, in the presence of caspase inhibitors, loss of mitochondrial cytochrome c does not commit NGF-deprived sympathetic neurons to die; instead, commitment to death is now concurrent with mitochondrial depolarization, which follows cytochrome c release by ~ 25–30 hours, culminating in a form of cell death, perhaps autophagic (Xue et al. 1999), lacking most, if not all, of the biochemical (e.g., Annexin V or TUNEL staining) and morphological hallmarks of apoptosis. Thus, caspase inhibition offers a finite, but real, therapeutic "window of opportunity" during which neurons can be rescued *if* the initiating death stimulus is aborted, or at least ameliorated, *and* a relatively normal, physiological milieu is restored. Furthermore, the duration of this therapeutic window may vary according to the cell type (Deshmukh and Johnson 2000), apoptotic insult, and stimulus strength. Taken together with other reports (Hara et al. 1997; Cheng et al. 1998; Ona et al. 1999; Li et al. 2000), these observations suggest that caspase inhibition as a clinical strategy for neurological disorders is most appropriate for acute conditions such as stroke. By contrast, for *chronic* neurodegenerative processes such as Alzheimer's Disease and Amyotrophic Lateral Sclerosis, modulation of the caspase checkpoint alone may be of limited efficacy in the absence of therapy designed to improve or reverse the underlying pathogenic mechanisms.

Acknowledgments. This work was supported by National Institutes of Health grants R37AG-12947 and RO1NS38651 (E.M.J.) and a Paralyzed Veterans of America Spinal Cord Research grant (M.D.). We thank A. Barbieri, L. Bernstein, P. Stahl, and M. Linder for assistance with subcellular fractionation and alkali extraction; P. Bouillet, J.M. Adams, and A. Strasser (The Walter and Eliza Hall Institute of Medical Research, Victoria, Australia) for *Bim*-deficient mice; S.J. Korsmeyer (Dana-Farber Cancer Institute) for *Bax*- and *Bad*-deficient mice; J.-C. Martinou (Serono Pharmaceutical Research Institute, Geneva, Switzerland) and S. Tonegawa (Massachusetts Institute of Technology) for the *Bcl-2* transgenic mice; M. Wallace and members of the Washington University Neuroscience Transgenic Core Facility for excellent mouse husbandry; P.A. Osborne for assistance with neuronal dissections; and M. Bloomgren for secretarial assistance.

References

Amarante-Mendes GP, Finucane DM, Martin SJ, Cotter TG, Salvesen GS, Green DR (1998a) Antiapoptotic oncogenes prevent caspase-dependent and independent commitment for cell death. Cell Death Differ 5:298–306

Amarante-Mendes GP, Naekyung Kim C, Liu L, Huang Y, Perkins CL, Green DR, Bhalla K (1998b) Bcr-Abl exerts its antiapoptotic effect against diverse apoptotic stimuli through blockage of mitochondrial release of cytochrome c and activation of caspase-3. Blood 91:1700–1705

Antonsson B, Conti F, Ciavatta A, Montessuit S, Lewis S, Martinou I, Bernasconi L, Bernard A, Mermod JJ, Mazzei G, Maundrell K, Gambale F, Sadoul R, Martinou J-C (1997) Inhibition of Bax channel-forming activity by Bcl-2. Science 277:370–372

Antonsson B, Montessuit S, Lauper S, Eskes R, Martinou J-C (2000) Bax oligomerization is required for channel-forming activity in liposomes and to trigger cytochrome c release from mitochondria. Biochem J 345:271–278

Bouillet P, Metcalf D, Huang DC, Tarlinton DM, Kay TW, Kontgen F, Adams JM, Strasser A (1999) Proapoptotic Bcl-2 relative Bim required for certain apoptotic responses, leukocyte homeostatis, and to preclude autoimmunity. Science 286:1735–1738

Bradham CA, Qian T, Streetz K, Trautwein C, Brenner DA, Lemasters JJ (1998) The mitochondrial permeability transition is required for tumor necrosis factor alpha-mediated apoptosis and cytochrome c release. Mol Cell Biol 18:6353–6364

Brunet CL, Gunby RH, Benson RSP, Hickman JA, Watson AJM, Brady G (1998) Commitment to cell death measured by loss of clonogenicity is separable from the appearance of apoptotic markers. Cell Death Differ 5:107–115

Cheng Y, Deshmukh M, D'Costa A, DeMaro JA, Gidday JM, Shah A, Sun Y, Jacquin MF, Johnson EM Jr, Holtzman DM (1998) Caspase inhibitor affords neuroprotection with delayed administration in a rat model of neonatal hypoxic-ischemic brain injury. J Clin Invest 101:1992–1999

Coughlin MD, Collins MB (1985) Nerve growth factor-independent development of embryonic mouse sympathetic neurons in dissociated cell culture. Dev Biol 110:392–401

Deckwerth TL, Johnson EM Jr. (1993) Temporal analysis of events associated with programmed cell death (apoptosis) of sympathetic neurons deprived of nerve growth factor. J Cell Biol 123:1207–1222

Deckwerth TL, Elliott JL, Knudson CM, Johnson EM Jr., Snider WD, Korsmeyer SJ (1996) Bax is required for neuronal death after trophic factor deprivation and during development. Neuron 17:401–411

Desagher S, Osen-Sand A, Nichols A, Eskes R, Montessuit S, Lauper S, Maundrell K, Antonsson B, Martinou J-C (1999) Bid-induced conformational change in Bax is responsible for mitochondrial cytochrome c release eduring apoptosis. J Cell Biol 144:891–901

Deshmukh M, Johnson EM Jr (1998) Evidence of a novel event during neuronal death: Development of competence-to-die in response to cytoplasmic cytochrome c. Neuron 21:695–705

Deshmukh M, Johnson EM Jr (2000) Caspase inhibition extends the commitment to neuronal death beyond cytochrome c release to the point of mitochondrial depolarization. J Cell Biol 150:131–143

Deshmukh M, Vasilakos J, Deckwerth TL, Lampe PA, Shivers BD, Johnson EM Jr (1996) Genetic and metabolic status of NGF-deprived sympathetic neurons saved by an inhibitor of ICE family proteases. J Cell Biol 135:1341–1354

Edwards SN, Buckmaster AE, Tolkovsky AM (1991) The death programme in cultured sympathetic neurones can be suppressed at the posttranslational level by nerve growth factor, cyclic AMP, and depolarization. J Neurochem 57:2140–2143

Ellis RE, Yuan JY, Horvitz HR (1991) Mechanisms and functions of cell death. Annu Rev Cell Biol 7:663–698

Eskes R, Antonsson B, Osensand A, Montessuit S, Richter C, Sadoul R, Mazzei G, Nichols A, Martinou J-C (1998) Bax-induced cytochrome c release from mitochondria is independent of the permeability transition pore but highly dependent on Mg^{2+} ions. J Cell Biol 143:217–224

Eskes R, Desagher S, Antonsson B, Martinou J-C (2000) Bid induces the oligomerization and insertion of Bax into the outer mitochondrial membrane. Mol Cell Biol 20:929–935

Estus S, Zaks WJ, Freeman RS, Gruda M, Bravo R, Johnson EM Jr (1994) Altered gene expression in neurons during programmed cell death: Identification of c-jun as necessary for neuronal apoptosis. J Cell Biol 127:1717–1727

Finucane DM, Bossy-Wetzel E, Waterhouse N, Cotter TG, Green DR (1999) Bax-induced caspase activation and apoptosis via cytochrome c release from mitochondria is inhibitable by Bcl-X$_L$. J Biol Chem 274:2225–2233

Freeman RS, Estus S, Johnson EM Jr. (1994) Analysis of cell cycle-related gene expression in postmitotic neurons: Selective induction of cyclin D1 during programmed cell death. Neuron 12:343–355

Gibson RM (1999) Caspase activation is downstream of commitment to apoptosis of Ntera-2 neuronal cells. Exp Cell Res 251:203–212

Goping IS, Gross A, Lavoie JN, Nguyen M, Jemmerson R, Roth K, Korsmeyer SJ, Shore GC (1998) Regulated targeting of bax to mitochondria. J Cell Biol 143:207–215

Green DR, Reed JC (1998) Mitochondria and apoptosis. Science 281:1309–1312

Greenlund LJ, Korsmeyer SJ, Johnson EM Jr (1995) Role of Bcl-2 in the survival and function of developing and mature sympathetic neurons. Neuron 15:649–661

Gross A, Jockel J, Wei MC, Korsmeyer SJ (1998) Enforced dimerization of Bax results in its translocation, mitochondrial dysfunction, and apoptosis. EMBO J 17:3878–3885

Hara H, Friedlander RM, Gagliardini V, Ayata C, Fink K, Huang Z, Shimizu-Sasamata M, Yuan J, Moskowitz MA (1997) Inhibition of interleukin 1 beta converting enzyme family proteases reduces ischemic and excitotoxic neuronal damage. Proc Natl Acad Sci USA 94:2007–2012

Heiskanen KM, Bhat MB, Wang HW, Ma J, Nieminen AL (1999) Mitochondrial depolarization accompanies cytochrome c release during apoptosis in PC6 cells. J Biol Chem 274:5654–5658

Hsu YT, Wolter KG, Youle RJ (1997) Cytosol-to-membrane redistribution of Bax and Bcl-X$_L$ during apoptosis. Proc Natl Acad Sci USA 94:3668–3672

Imaizumi K, Tsuda M, Imai Y, Wanaka A, Takagi T, Tohyama M (1997) Molecular cloning of a novel polypeptide, dp5, induced during programmed neuronal death. J Biol Chem 272:18842–18848

Inohara N, Ding LY, Chen S, Nunez G (1997) Harakiri, a novel regulator of cell death, encodes a protein that activates apoptosis and interacts selectively with survival-promoting proteins Bcl-2 and Bcl-X$_L$. EMBO J 16:1686–1694

Johnson EM Jr, Chang JY, Koike T, Martin DP (1989) Why do neurons die when deprived of trophic factor? Neurobiol Aging 10:549–552

Jürgensmeier JM, Xie Z, Deveraux Q, Ellerby L, Bredesen D, Reed JC (1998) Bax directly induces release of cytochrome c from isolated mitochondria. Proc Natl Acad Sci USA 95:4997–5002

Kerr JF, Wyllie AH, Currie AR (1972) Apoptosis: a basic biological phenomenon with wide-ranging implications in tissue kinetics. Br J Cancer 26:239–257

Kluck RM, Bossywetzel E, Green DR, Newmeyer DD (1997) The release of cytochrome c from mitochondria: A primary site for Bcl-2 regulation of apoptosis. Science 275:1132–1136

Knudson CM, Korsmeyer SJ (1997) Bcl-2 and Bax function independently to regulate cell death. Nat Genetics 16:358–363

Knudson CM, Tung KS, Tourtellotte WG, Brown GA, Korsmeyer SJ (1995) Bax-deficient mice with lymphoid hyperplasia and male germ cell death. Science 270:96–99

Koike T, Martin DP, Johnson EM Jr (1989) Role of Ca^{2+} channels in the ability of membrane depolarization to prevent neuronal death induced by trophic-factor deprivation: Evidence that levels of internal Ca^{2+} determine nerve growth factor dependence of sympathetic ganglion cells. Proc Natl Acad Sci USA 86:6421–6425

Kroemer G, Reed JC (2000) Mitochondrial control of cell death. Nat Med 6:513–519

Levi-Montalcini R, Booker B (1960) Destruction of the sympathetic ganglia in mammals by an antiserum to the nerve growth-promoting factor. Proc Natl Acad Sci USA 46:384–391

Li HL, Zhu H, Xu CJ, Yuan JY (1998) Cleavage of Bid by caspase-8 mediates the mitochondrial damage in the Fas pathway of apoptosis. Cell 94:491–501

Li MW, Ona VO, Guegan C, Chen MH, Jackson-Lewis V, Andrews LJ, Olszewski AJ, Stieg PE, Lee JP, Przedborski S, Friedlander RM (2000) Functional role of caspase-1 and caspase-3 in an ALS transgenic mouse model. Science 288:335–339

Li P, Nijhawan D, Budihardjo I, Srinivasula SM, Ahmad M, Alnemri ES, Wang X (1997) Cytochrome c and dATP-dependent formation of Apaf-1/caspase-9 complex initiates an apoptotic protease cascade. Cell 91:479–489

Liu X, Kim CN, Yang J, Jemmerson R, Wang X (1996) Induction of apoptotic program in cell-free extracts: Requirement for dATP and cytochrome c. Cell 86:147–157

Luo X, Budihardjo I, Zou H, Slaughter C, Wang XD (1998) Bid, a Bcl-2 interacting protein, mediates cytochrome c release from mitochondria in response to activation of cell surface death receptors. Cell 94:481–490

Martin DP, Schmidt RE, DiStefano PS, Lowry OH, Carter JG, Johnson EM Jr (1988) Inhibitors of protein synthesis and RNA synthesis prevent neuronal death caused by nerve growth factor deprivation. J Cell Biol 106:829–844

Martinou I, Desagher S, Eskes R, Antonsson B, Andre E, Fakan S, Martinou J-C (1999) The release of cytochrome c from mitochondria during apoptosis of NGF-deprived sympathetic neurons is a reversible event. J Cell Biol 144:883–889

Marzo I, Brenner C, Zamzami N, Jürgensmeier JM, Susin SA, Vieira HLA, Prevost MC, Xie ZH, Matsuyama S, Reed JC, Kroemer G (1998) Bax and adenine nucleotide translocator cooperate in the mitochondrial control of apoptosis. Science 281:2027–2031

McCarthy MJ, Rubin LL, Philpott KL (1997) Involvement of caspases in sympathetic neuron apoptosis. J Cell Sci 110:2165–2173

Metzstein MM, Stanfield GM, Horvitz HR (1998) Genetics of programmed cell death in C. elegans: Past, present and future. Trend Genet 14:410–416

Miller TM, Moulder KL, Knudson CM, Creedon DJ, Deshmukh M, Korsmeyer SJ, Johnson EM Jr (1997) Bax deletion further orders the cell death pathway in cerebellar granule cells and suggests a caspase-independent pathway to cell death. J Cell Biol 139:205–217

Minn AJ, Velez P, Schendel SL, Liang H, Muchmore SW, Fesik SW, Fill M, Thompson CB (1997) Bcl-X_L forms an ion channel in synthetic lipid membranes. Nature 385:353–357

Muchmore SW, Sattler M, Liang H, Meadows RP, Harlan JE, Yoon HS, Nettesheim D, Chang BS, Thompson CB, Wong SL, Ng SL, Fesik SW (1996) X-ray and NMR structure of human Bcl-X_L, an inhibitor of programmed cell death. Nature 381:335–341

Narita M, Shimizu S, Ito T, Chittenden T, Lutz RJ, Matsuda H, Tsujimoto Y (1998) Bax interacts with the permeability transition pore to induce permeability transition and cytochrome c release in isolated mitochondria. Proc Natl Acad Sci USA 95:14681–14686

Neame SJ, Rubin LL, Philpott KL (1998) Blocking cytochrome c activity within intact neurons inhibits apoptosis. J Cell Biol 142:1583–1593

O'Connor L, Strasser A, O'Reilly LA, Hausmann G, Adams JM, Cory S, Huang DC (1998) Bim: a novel member of the Bcl-2 family that promotes apoptosis. EMBO J 17:384–395

Ohta T, Kinoshita T, Naito M, Nozaki T, Masutani M, Tsuruo T, Miyajima A (1997) Requirement of the caspase-3/CPP32 protease cascade for apoptotic death following cytokine deprivation in hematopoietic cells. J Biol Chem 272:23111–23116

Oltvai ZN, Korsmeyer SJ (1994) Checkpoints of dueling dimers foil death wishes. Cell 79:189–192

Oltvai ZN, Milliman CL, Korsmeyer SJ (1993) Bcl-2 heterodimerizes in vivo with a conserved homolog, Bax, that accelerates programmed cell death. Cell 74:609–619

Ona VO, Li M, Vonsattel JP, Andrews LJ, Khan SQ, Chung WM, Frey AS, Menon AS, Li XJ, Stieg PE, Yuan J, Penney JB, Young AB, Cha JH, Friedlander RM (1999) Inhibition of caspase-1 slows disease progression in a mouse model of Huntington's disease. Nature 399:263–267

Oppenheim RW (1991) Cell death during development of the nervous system. Ann Rev Neurosci 14:453–501

O'Reilly LA, Cullen L, Moriishi K, O'Connor L, Huang DC, Strasser A (1998) Rapid hybridoma screening method for the identification of monoclonal antibodies to low-abundance cytoplasmic proteins. Biotech 25:824–830

O'Reilly LA, Cullen L, Visvader J, Lindeman GJ, Print C, Bath ML, Huang DC, Strasser A (2000) The proapoptotic BH3-only protein Bim is expressed in hematopoietic, epithelial, neuronal, and germ cells. Am J Pathol 157:449–461

Putcha GV, Deshmukh M, Johnson EM Jr (1999) Bax translocation is a critical event in neuronal apoptosis: Regulation by neuroprotectants, Bcl-2, and caspases. J Neurosci 19:7476–7485

Putcha GV, Deshmukh M, Johnson EM Jr (2000) Inhibition of apoptotic signaling cascades causes loss of trophic factor dependence during neuronal maturation. J Cell Biol 149:1011–1018

Puthalakath H, Huang DC, O'Reilly LA, King SM, Strasser A (1999) The proapoptotic activity of the Bcl-2 family member Bim is regulated by interaction with the dynein motor complex. Molecular Cell 3:287–296

Rosse T, Olivier R, Monney L, Rager M, Conus S, Fellay I, Jansen B, Borner C (1998) Bcl-2 prolongs cell survival after Bax-induced release of cytochrome c. Nature 391:496–499

Rydel RE, Greene LA (1988) cAMP analogs promote survival and neurite outgrowth in cultures of rat sympathetic and sensory neurons independently of nerve growth factor. Proc Natl Acad Sci USA 85:1257–1261

Saikumar P, Dong Z, Patel Y, Hall K, Hopfer U, Weinberg JM, Venkatachalam MA (1998) Role of hypoxia-induced Bax translocation and cytochrome c release in reoxygenation injury. Oncogene 17:3401–3415

Saito M, Korsmeyer SJ, Schlesinger PH (2000) Bax-dependent transport of cytochrome c reconstituted in pure liposomes. Nat Cell Biol 2:553–555

Schendel SL, Xie Z, Montal MO, Matsuyama S, Montal M, Reed JC (1997) Channel formation by antiapoptotic protein Bcl-2. Proc Natl Acad Sci USA 94:5113–5118

Schindler KS, Latham CB, Roth KA (1997) Bax deficiency prevents the increased cell death of immature neurons in Bcl-X-deficient mice. J Neurosci 17:3112–3119

Schlesinger PH, Gross A, Yin XM, Yamamoto K, Saito M, Waksman G, Korsmeyer SJ (1997) Comparison of the ion channel characteristics of proapoptotic Bax and antiapoptotic Bcl-2. Proc Natl Acad Sci USA 94:11357–11362

Stennicke HR, Salvesen GS (2000) Caspases: controlling intracellular signals by protease zymogen activation. Biochim Biophys Acta 1477:299–306

Thompson CB (1995) Apoptosis in the pathogenesis and treatment of disease. Science 267:1456–1462

Troy CM, Stefanis L, Prochiantz A, Greene LA, Shelanski ML (1996) The contrasting roles of Ice family proteases and interleukin-1-β in apoptosis induced by trophic factor withdrawal and by copper/zinc superoxide dismutase down-regulation. Proc Natl Acad Sci USA 93:5635–5640

Werth JL, Deshmukh M, Cocabo J, Johnson EM, Rothman SM (2000) Reversible physiological alterations in sympathetic neurons deprived of NGF but protected from apoptosis by caspase inhibition or Bax deletion. Exp Neurol 161:203–211

White FA, Kellerpeck CR, Knudson CM, Korsmeyer SJ, Snider WD (1998) Widespread elimination of naturally occurring neuronal death in Bax-deficient mice. J Neurosci 18:1428–1439

Wolter KG, Hsu YT, Smith CL, Nechushtan A, Xi XG, Youle RJ (1997) Movement of Bax from the cytosol to mitochondria during apoptosis. J Cell Biol 139:1281–1292

Xiang JL, Chao DT, Korsmeyer SJ (1996) Bax-induced cell death may not require interleukin 1-beta-converting enzyme-like proteases. Proc Natl Acad Sci USA 93:14559–14563

Xue L, Fletcher GC, Tolkovsky AM (1999) Autophagy is activated by apoptotic signalling in sympathetic neurons: An alternative mechanism of death execution. Mol Cell Neurosci 14:180–198

Yang J, Liu XS, Bhalla K, Kim CN, Ibrado AM, Cai JY, Peng TI, Jones DP, Wang XD (1997) Prevention of apoptosis by Bcl-2: release of cytochrome c from mitochondria blocked. Science 275:1129–1132

Yin XM, Wang K, Gross A, Zhao Y, Zinkel S, Klocke B, Roth KA, Korsmeyer SJ (1999) Bid-deficient mice are resistant to Fas-induced hepatocellular apoptosis. Nature 400:886–891

Zou H, Henzel WJ, Liu X, Lutschg A, Wang X (1997) Apaf-1, a human protein homologous to C. elegans CED-4, participates in cytochrome c-dependent activation of caspase-3. Cell 90:405–413

Normal Programmed Cell Death of Developing Avian and Mammalian Neurons Following Inhibition or Genetic Deletion of Caspases

R. W. Oppenheim, C.-Y. Kuan, D. Prevette, P. Rakic, and H. Yaginuma

Summary

Caspase inhibitors fail to prevent the programmed cell death (PCD) of developing motoneurons (MNs) in the chick embryo in ovo, and genetic deletion of upstream (caspase-9) or downstream (caspase-3) caspase family members in mice does not prevent the PCD of developing MNs or other populations of target-dependent post-mitotic neurons. Neurons undergoing PCD in the absence of caspases exhibit a non-apoptotic, often TUNEL-negative, mode of morphological degeneration. Loss of caspase activity may delay but does not prevent the quantitatively normal occurrence of neuronal PCD.

Introduction

Cysteine proteases comprising the caspase family are considered to be among the most highly conserved molecules involved in the apoptosis and PCD of developing cells, being expressed in animals as diverse as worms and humans (Cryus and Yuan 1998; Li and Yuan 1999). They are one of the major classes of pro-apoptotic molecules and are thought to be essential for many of the degradative events that occur during the cell death process. By cleaving specific substrates in the nucleus and cytoplasm, caspases are responsible for many of the biochemical and cytological changes that define apoptotic PCD.

Although the entire apoptotic pathway has not yet been completely defined biochemically, many of the essential steps are rapidly being elucidated. For example, the role of caspase-3 (casp3) is thought to involve a site in the pathway downstream of other caspases. Casp3 can be activated by caspase-8 in the process of Fas-mediated PCD (Nagata 1997) and by casp9, in combination with cytochrome-c released from mitochondria, and Apaf-1 (Liu et al. 1997). Activated casp3 cleaves ICAD/DFF45 (Liu et al. 1997; Sakahira et al. 1998), leading to the activation of CAD/DFF40 that results in DNA fragmentation (Enari et al. 1998; Liu et al. 1998). Casp3 also activates casp6, which then acts to cleave nuclear proteins and mediates the shrinkage and fragmentation of the nucleus (Hirata et al. 1998; Kawahara et al. 1998).

Previous in vitro studies have shown that various kinds of stimuli that normally induce apoptosis, chromatin condensation and DNA degradation

Henderson/Green/Mariani/Christen (Eds.)
Neuronal Death by Accident or by Design
© Springer-Verlag Berlin Heidelberg 2001

Table 1. Some features of motoneuron cell death in the chick embryo

Feature	Early cervical MN death	Death of other MNs
1) Timing	Rapid, E4–E5	Prolonged, E6–E12
2) Target-dependent	no	yes
3) Trophic factors	GDNF[a] family members only	MEX, GDNF, CNTF, IGF, BDNF, BMP, CT-1, HGF
4) Mode of PCD	Apoptotic	Apoptotic
5) Caspase-dependent	?	yes
6) Post-mitotic	yes	yes
7) Rescued by Cyclohex. or Act-D	yes	yes

[a] glia-derived neurotrophic factor

fail to do so in neuronal and non-neuronal cells that are casp3 deficient or in which casp3 or other caspases are inhibited (Bortner and Cidlowski 1999; Cregan et al. 1999; D'Mello et al. 1999; Ferrer 1999; Jänicke et al. 1998a,b; Keramaris et al. 2000; Stefanis et al. 1998, 1999; Tanabe et al. 1999; Woo et al. 1998; Xue et al. 1999; Zheng et al. 1998). Accordingly, casp3, and by extrapolation, casp9, which is thought to be upstream of, and required for, casp3 activation, both appear to be mainly required for the nuclear changes that occur during apoptosis. By contrast, in vivo sutdies of casp3- and casp9-deficient mice indicate that these proteases are, in fact, essential for both the nuclear changes and the normal PCD of many developing neurons (Kuida et al. 1996, 1998; Hakem et al. 1998; Roth et al. 2000). However, these studies focused almost exclusively on mitotically active or immature neurons in the forebrain that undergo normal cell death during early developmental stages prior to the formation of synaptic connections.

Because the regulation of neuronal PCD in vitro, or in populations in vivo that undergo PCD prior to the formation of synaptic connections, may differ from the well known target-dependent type of naturally occurring neuronal death that occurs when synaptic connections are being formed (Oppenheim 1999), these in vitro and in vivo studies on the role of casp3 and casp9 may not accurately reflect the in vivo role of these proteases in this type of PCD. To examine this possibility, we have focused on a population of MNs in the chick embryo cervical spinal cord that differ in several respects from other spinal and cranial MNs, most notably by their early, rapid and extensive PCD (Yaginuma et al. 1996; Table 1). Additionally, we have examined PCD of several populations of developing post-mitotic neurons in mice in which either casp3 or casp9 has been genetically deleted (Oppenheim et al. 2001).

Cervical MNs in the Chick Embryo

In most population of developing vertebrate MNs, approximately one-half of the initially generated post-mitotic cells degenerate by an apoptotic-like pathway over a period of several days (Oppenheim 1991). By contrast, approximately 70% of chick cervical MNs undergo PCD over a brief 24-hour period between embryonic day (E) 4 and E5 (Yaginuma et al. 1996). Cervical MN death is apoptotic and involves DNA fragmentation detected by TUNEL labeling (Fig. 1). To examine the role of caspases in this early, massive PCD of cervical MNs, we measured caspase activity in the spinal cord and analyzed the effects of in ovo treatment with caspase inhibitors on PCD (Shiraiwa et al., in press).

There was a significant and selective increase in caspase activity (casp3) in the ventral spinal cord of cervical segments on E4 but not in thoracic segments where MN PCD occurs later, between E6 and E10 (Fig. 2). Following 12 hours of treatment with the caspase inhibitor AC-DEVD-CHO, the nuclei of degenerating MNs exhibited only a very moderate condensation without TUNEL labeling when examined in the light microscope (Fig. 1). By contrast, control MNs exhibited typical condensed pyknotic nuclei that were positive for TUNEL labeling (Fig. 1). When both types of degenerating MNs were counted, no significant differences were found between control vs DEVD-treated embryos (Fig. 3), suggesting that despite caspase inhibition, MNs were still undergoing PCD. Ultrastructural examination revealed that MNs in DEVD-treated embryos exhibit a non-apoptotic morphology with few nuclear changes but with vacuolar degenerative changes in the cytoplasm not observed in control MNs (Fig. 1). Because the degenerating MNs in DEVD-treated embryos were often observed in the electron microscope to be in the process of being phagocytozed by neighboring cells or macrophages (not shown), they were obviously dying cells. This finding was confirmed by the quantification of MN numbers at the end of the normal cell death period on E5.5 (Fig. 4), indicating that comparable numbers of MNs had been lost during the preceding 24–36 hours (E4–E5.5) in both control and DEVD-treated embryos.

Surprisingly, following more prolonged DEVD treatment (24 vs 12 hours), there were increased numbers of degenerating MNs on E5 compared to controls (Fig. 3), and now most of these dying MNs were TUNEL positive (Fig. 3). Because cell counts of surviving MNs on E5 (see Fig. 4) indicate that the amount of cell death was similar in control vs DEVD-treated embryos, caspase inhibition appears to only delay but not prevent the PCD of MNs. Futhermore, the appearance of TUNEL labeling in a significant proportion of DEVD-treated MNs on E5 vs E4.5 suggests that the nuclear changes detected by TUNEL can occur even in the absence of those caspases that are inhibited by DEVD treatment.

In contrast to the effects of DEVD treatment, treatment of embryos with the more wide spectrum, pan-caspase inhibitor, boc-aspartyl (OMe)-fluoromethylketone (BAF), revealed both similarities and differences compared to

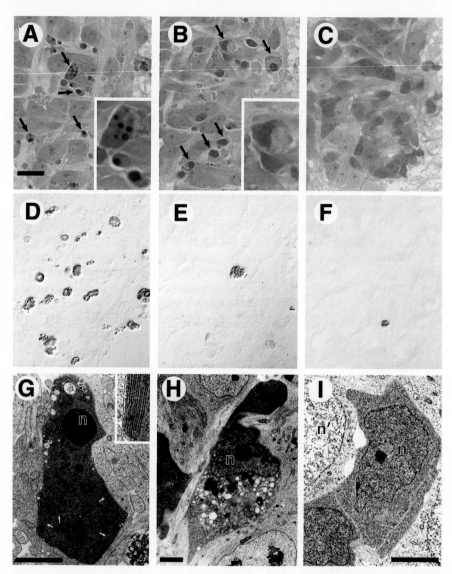

Fig. 1. Light and electron photomicrographs of the ventral horn of the cervical spinal cord of the E4.5 chick embryos following treatment with caspase inhibitors for 12 hours. **A, D, G,** control. **B, E, H,** treatment with 400 µg of Ac-DEVD-CHO. **C, F, I,** treatment with 200 µg of BAF. **A–C,** hematoxylin and eosin staining. **D–F,** TUNEL staining. **G–I,** electron micrographs. Arrows in **A, B** indicated examples of pyknotic neurons. Insets in **A** and **B** show higher magnification of the typical morphology of pyknotic neurons. Note the different morphologies of pyknotic cells in controls vs Ac-DEVD-CHO (**A, B**). Few pyknotic cells were seen in the embryos treated with Ac-DEVD-CHO or BAF (**E, F**). **G,** typical example of the type of degeneration found in a control embryo. Condensation of chromatin (**n**), aggregated ribosomes (arrows and inset) and increased electron density of the cytoplasm are distinct. **H,** aberrantly degenerating cells in embryos treated with Ac-DEVD-CHO for 12 hours. Note that although the cytoplasm exhibited degenerative changes, extreme shrinkage and fragmentation of nucleus (**n**) did not occur. Arrowheads in **G, H** indicate clear vacuoles that include debris from cell organelles. **I,** cells in embryos treated with BAF for 12 hours. There were many cells with slightly condensed cytoplasm and nuclei. Compare the slightly condensed and irregular shaped cell nucleus (**n**) with the nucleus of adjacent cell (**n′**) that is apparently normal. Scale bar in **A–F**, 10 µg; scale bars in **G–I**, 2 µg

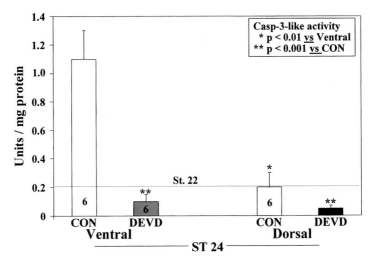

Fig. 2. Casp3-like activity in the cervical spinal cord as assayed using fluorogenic substrates to estimate enzyme activity. Data reflect three independent experiments (n = 5 per group) for each value. The horizontal line represents enzyme activity in whole spinal cord at st.22 prior to the onset of cervical MN death, CON, control

the effects of DEVD. Following 12 hours of treatment with BAF, there is a significant reduction in the numbers of degenerating cells if one counts both typical pyknotic MNs as well as those cells exhibiting the kind of atypical morphology observed following DEVD treatment (Fig. 3). However, there were also many cells in BAF-treated embryos that had irregularly shaped nuclei, increased eosinophilic staining when observed in the light microscope and were TUNEL negative (Fig. 1). These cells differ from the degenerating cells seen following treatment with DEVD. Ultrastructural observations suggest that these BAF-treated MNs are in an early stage of degeneration, as indicated by the presence of electron-dense cytoplasm and nucleus. Accordingly, if one includes these cells in the counts of degenerating MN's, then again, similar to DEVD treatment, the numbers on E4.5 are comparable to controls (not shown). Prolonged treatment with BAF for 24 hours results in increased numbers of normally pyknotic, TUNEL-positive cells on E5, and counts of surviving MNs on E5 indicate that BAF treatment delays but does not prevent the loss of normal numbers of MNs by PCD (Fig. 4). The different, less extensive nuclear and cytoplasmic changes in BAF-treated MNs at 12 hours (E4.5), compared to DEVD, suggest that other upstream caspases inhibited by BAF, but not by DEVD, are involved in the more extensive degenerative changes observed at 12 hours following DEVD treatment. However, both DEVD and BAF only delay but do not prevent normal PCD, suggesting that, in the absence of caspase activity, these MNs may undergo a caspase-independent form of cell death with reduced or absent TUNEL labeling and with a distinct non-apoptotic morphology.

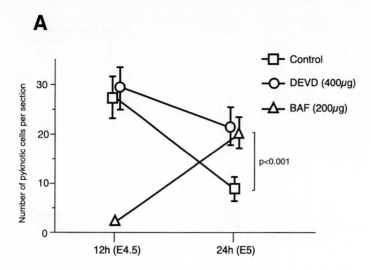

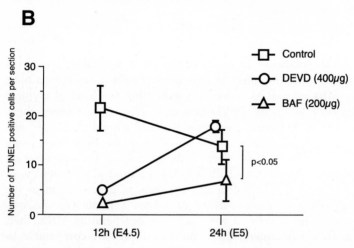

Fig. 3. Quantitative analyses of the number of pyknotic cells (**A**) and TUNEL-positive cells (**B**) in the ventral horn region of the cervical cord of chick embryos treated with Ac-DEVD-CHO or BAF for 12 hours and 24 hours. The results are presented as the mean number of cells ($\pm$SD) per section through the C10 segment

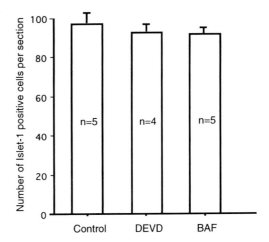

Fig. 4. Quantitative analyses of the numbers of healthy motoneurons that exhibited Islet-1 immunolabeling on E5 following treatment with Ac-DEVD-CHO or BAF for 24 hours. The results are presented as the mean number of Islet-1-positive cells (±SD) per 7-μm thick section through the C8–11 segments

Neuronal PCD in Casp3 and Casp9 Knock-out Mice

Because the genetic deletion of casp3 (Kuida et al. 1996; Woo et al. 1998), casp9 (Hakem et al. 1998; Kuida et al. 1998) and Apaf-1 (Yoshida et al. 1998; Cecconi et al. 1998) results in a severe and distinct CNS phenotype that has been attributed, in part, to reduced PCD, these genes have been suggested to comprise an obligatory linear pathway for some kinds of neuronal PCD (Kuan et al. 2000). Because the PCD of many non-neuronal cell populations is relatively normal in these knock-outs (KO) (e.g., interdigital cell death), it has been argued that this gene pathway is especially important for neuronal development. However, the evidence for this argument is based almost entirely on the analysis of early cell death in the forebrain prior to the onset of neuronal connectivity. Because the lion's share of neuronal PCD in the vertebrate nervous system occurs later, when neurons are establishing synaptic connections with targets and afferents (Oppenheim 1991), we have asked whether cell death during this later developmental period also requires the casp9 → Apaf-1 → casp3 pathway (Oppenheim et al. 2001).

Although there is an apparent marked reduction in the number of TUNEL-positive cells in the spinal cord and brainstem of casp9- (Kuida et al. 1998) and casp3- (own data, not shown) deleted embryos, many "pyknotic" neurons could be observed in Nissl-stained sections (Fig. 5). When quantified, no significant differences in the number of degenerating cells between wild-type and heterozygous and homozygous casp3- or casp9-deleted embryos were detected (Fig. 6). When examined in either the light or electron microscope, dying neurons in the KO (–/–) embryos exhibited a morphology distinct from that of either wild-type or heterozygote (+/–) embryos (Fig. 5). In contrast to the typical pyknotic, apoptotic morphology of dying control or heterozygote neurons, degenerating neurons in the KO embryos were intact, but reduced in size, exhibited less nuclear condensation and were darker. In

Casp3 +/-   Casp3 -/-

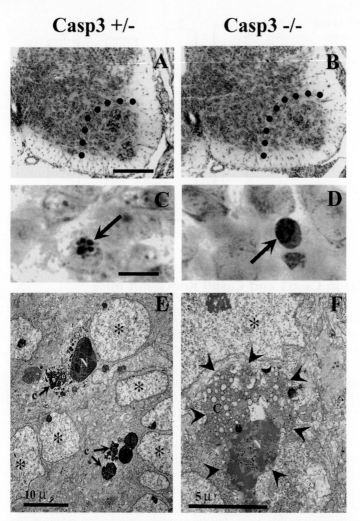

Fig. 5. A, B, spinal cord sections (lumbar, E15.5) and the morphology (C–F) of degenerating spinal motoneurons in casp3 hetero- and homozygote embryos. Although not shown, casp9 +/- and -/- embryos exhibit a similar morphology and the neuronal morphology of wild-type control embryos (+/+) is indistinguishable from casp3 and casp9 heterozygotes. In the light microscope, dying motoneurons in KO embryos (**D**) exhibit less chromatin condensation and darker cytoplasm than controls (**C**). Ultrastructurally, striking features of dying motoneurons in KO embryos are numerous cytoplasmic vacuoles (**F**). The dotted line in **A, B** delineates the border of the ventral horn. In **F**, arrowheads outline the soma. Asterisks in **E** indicate nuclei of healthy surviving neurons. N, nucleus; C, cytoplasmic debris

the electron microscope, neurons in KO animals exhibit cytoplasmic changes (e.g., vacuoles, dilated mitochondria) rarely observed in the degenerating neurons, from control or heterozygote embryos. Although differing in some respects, these degenerating mouse neurons nonetheless resemble the dying

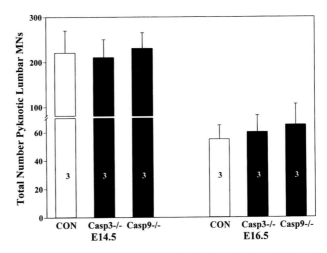

Fig. 6. The number (mean ±SD) of degenerating motoneurons in the lumbar spinal cord of casp3 KO, casp9 KO and control (CON) embryos on E14.5 and E16.5

neurons observed in the spinal cord of DEVD- and BAF-treated chick embryos described above.

If these morphologically aberrant neurons are, in fact, undergoing PCD, then one would expect that the number of surviving neurons at the end of the normal cell death period would be comparable in wild-type, heterozygote and homozygote animals. In fact, this is the case (Tables 2, 3). Because most casp9 KO embryos do not survice past E17, our analysis of surviving neurons was restricted to populations in which substantial normal PCD occurs prior to E17 (MNs and DRG). By contrast, more neuronal populations could be assessed in casp3 KO animals, because many of these animals survive after birth. In all populations that we examined, which includes spinal and cranial MNs, sensory and sympathetic neurons and spinal interneurons, we could detect no differences in the number of surviving neurons in control vs KO animals. In at least one population (lumbar MNs), cell numbers were also comparable at the beginning of the cell death period on E13–E14 in control and KO (casp3) embryos (not shown), indicating that early events such as proliferation and migration were also not perturbed by the absence of caspase activity. Finally, although all casp3 and casp9 KO animals used in our analysis were confirmed to exhibit forebrain hyperplasia, the histology of the brainstem and spinal cord appeared normal (Figs. 5, 7).

In summary, our analysis of PCD of several populations of post-mitotic neurons in casp3- and casp9-depleted mice shows that, despite the absence or reduction of TUNEL labeling, normal numbers of these neurons undergo cell death. The morphology of degenerating neurons, however, is non-apoptotic and ultrastructurally distinct from degenerating neurons in control animals.

Table 2. Neuronal numbers (mean ± SD) in postnatal (P) caspase-3-deficient mice

	P0–P2		P8–P20[a]		P42	
	+/+	–/–	+/+	–/–	+/+	–/–
Lumbar MNs	1738 ± 179	1895 ± 220	1806 ± 173	1788 ± 205	1650	1687
	n = 8	n = 8	n = 7	n = 6	n = 2	n = 2
Brachial MNs	–	–	1710 ± 98	1770 ± 113	–	–
			n = 6	n = 6		
Facial MNs	3145 ± 196	3220 ± 260	2914 ± 200	2980 ± 221	–	–
	n = 3	n = 3	n = 5	n = 4		
Hypoglossal MNs	–	–	1434 ± 81	1273 ± 113	–	–
			n = 5	n = 5		
Trigeminal MNs	–	–	978 ± 61	936 ± 54	–	–
			n = 4	n = 4		
Abducens MNs	–	–	174 ± 13	178 ± 18	–	–
			n = 5	n = 5		
Oculomotor MNs	–	–	531 ± 40	542 ± 33	–	–
			n = 3	n = 3		
Interneurons (lumbar)	–	–	–	–	20 280	22 317
					n = 2	n = 2
DRG (L4)	–	–	4455 ± 210	4185 ± 237	5691	5359
			n = 4	n = 4	n = 2	n = 2
SCG[b]	–	–	12 180 ± 747	12 991 ± 1324	–	–
			n = 4	n = 4		

[a] Includes data from P8, P10 and P20.
[b] Data from P8 only

Table 3. Neuronal numbers (mean ± SD) in caspase-3- and caspase-9-deficient mice on E16.5

	Lumbar MNs		Facial MNs		Hypoglossal MNs	
	casp9	casp3	casp9	casp3	casp9	casp3
Control (+/+, +/–)	2372 ± 169	2224 ± 200	3985 ± 671	4000 ± 480	1790 ± 465	1610 ± 360
	n = 6	n = 5	n = 4	n = 5	n = 4	n = 5
Mutant (–/–)	2421 ± 183	2430	3614 ± 510	3600	1715 ± 410	1550
	n = 5	n = 2	n = 3	n = 2	n = 4	n = 2

Conclusion

Although it is generally believed that the casp9 → Apaf-1 → casp3 execution pathway is essential for the PCD of many population of develpoping cells, especially neurons, the present results indicate that inhibition of caspases (chick) or genetic deletion of either casp3 or casp9 (mice) does not prevent the normal loss of several types of neurons. Because a similar analysis has not been carried out with Apaf-1 mutants, it is not kown whether neuronal PCD also occurs normally in these mice. However, based on the report that the spinal cord and brainstem of these mutants appear normal, it seems

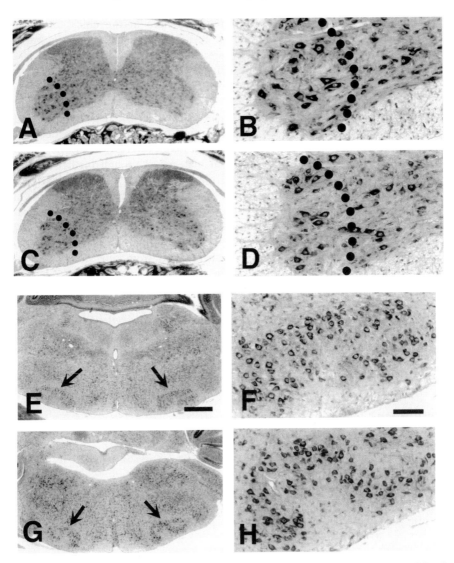

Fig. 7. Transverse sections of the lumbar spinal cord (**A–D**) and brainstem (**E–H**; level of facial motor nucleus, arrows in **E, G**) of P9 casp3 KO (**C, D, G, H**) and control (**A, B, E, F**) animals. Enlargements show the ventral horn (**B, D**) and facial nucleus (**F, H**). Scale bar in E = 100 μm (same for A, C, G) and in F = 60 μm (same for B, D, H)

likely that Apaf-1 may also be dispensable for the PCD of the same populations of post-mitotic neurons analyzed here (Cecconi et al. 1998).

Because neither the results from the chick embryo nor the analysis of casp3 or casp9 KO mice can exclude the possibility that other caspases not affected by these perturbations may mediate PCD, we cannot determine whether cell death in these situations is truly caspase-independent. However, the normal occurrence of cell death reported here in the chick following treatment with the pan-caspase inhibitor BAF and similar results in cultured sympathetic neurons (Xue et al. 1999) are consistent with the possibility of a caspase-independent pathway. Some evidence indicates that, in the absence of caspase activation, cells may undergo PCD by a pathway involving lysosomal proteases and autophagy (Ishahara et al. 1999; Xue et al. 1999; McIlroy et al. 2000). However, this is unlikely to be the case for the degenerating neurons we observed following caspase inhibition or deletion. The vesicles or vacuoles that are characteristic of dying neurons in both chick and mouse in this situation appear empty of lysosomal material. Although further studies are needed, the dying neurons that we observed appear to more closely fit the type 3B "cytoplasmic" mode of cell death described by Clarke (1990), in which there is modest chromatin condensation and dilation and vascuolization of cytoplasmic organelles (also see Chu-Wang and Oppenheim 1978; Pilar and Landmesser 1976).

Finally, an important cautionary note derived from our studies is that TUNEL labeling by itself is not always a reliable index of the occurrence or extent of neuronal PCD. As we have shown, the absence of, or a reduction in, TUNEL labeling does not necessarily reflect reduced PCD in the nervous system. The same thing appears to be true for many other cell types and situations (e.g., Chautan et al. 1999; Stefanis et al. 1999; McCarthy et al. 1997; Tanabe et al. 1999).

References

Bortner CD, Cidlowski JA (1999) Caspase independent/dependent regulation of K^+, cell shrinkage and mitochondrial membrane potential during lymphocyte apoptosis. J Biol Chem 274:21953–21962

Cecconi F, Alvarez-Bolado G, Meyer BI, Roth KA, Gruss P (1998) Apaf-1 (CED-4 homolog) regulates programmed cell death in mammalian development. Cell 94:727–737

Chautan M, Chazal G, Cecconi F, Gruss P, Golstein P (1999) Interdigital cell death can occur through a necrotic and caspase-independent pathway. Curr Biol 9:967–970

Chu-Wang IW, Oppenheim RW (1978) Cell death of motoneurons in the chick embryo spinal cord. J Comp Neurol 177:33–58

Clarke PGH (1990) Developmental cell death: morphological diversity and multiple mechanisms. Anat Embryol 181:195–213

Cregan SP, MacLaurin JG, Craig CG, Robertson GS, Nicholson DW, Park DS, Slack RS (1999) Bax-dependent caspase-3 activation is a key determinant in p53-induced apoptosis in neurons. J Neurosci 19:7860–7869

Cryus V, Yuan J (1998) Proteases to die for. Genes Dev 12:1551–1570

D'Mello SR, Kuan C-Y, Flavell R, Rakic P (1999) Caspase-3 is required for apoptosis-associated DNA fragmentation but not for cell death in neurons deprived of potassium. Soc Neurosci Abst 25:551

Enari M, Sakahira H, Yokoyama H, Okawa K, Iwamatsu A, Nagata S (1998) A caspase activated DNase that degrades DNA during apoptosis and its inhibitor ICAD. Nature 391:43–50

Ferrer I (1999) Role of caspases in ionizing radiation-induced apoptosis in the developing cerebellum. J Neurobiol 41:549–558

Hakem R, Hakem A, Duncan GS, Henderson JT, Woo M, Soengas M, Elia A, de la Pompa JL, Kagi D, Khoo W, Potter J, Yoshida R, Kaufman SA, Lowe SW, Penninger JM, Mak TW (1998) Differential requirement for caspase 9 in apoptotic pathways in vivo. Cell 94:339–352

Hirata H, Takahashi A, Kobayashi S, Yonehara S, Sawai H, Okazaki T, Yamamoto K, Sasasda M (1998) Caspases are activated in a branched protease cascade and control distinct downstream processes in Fas-induced apoptosis. J Exp Med 187:587–600

Isahara K, Ohsawa Y, Kanamori S, Shibata M, Waguri S, Sato N, Gotow T, Watanabe T, Momoi T, Urase K, Kominami E, Uchiyama Y (1999) Regulation of a novel pathway for cell death by lysosomal aspartic and cysteine proteinases. Neuroscience 91:233–249

Jänicke RU, Sprengart ML, Wati MR, Porter AG (1998a) Caspase-3 is required for DNA fragmentation and morphological changes associated with apoptosis. J Biol Chem 273:9357–9360

Jänicke RU, Ng P, Sprengart ML, Porter AG (1998b) Caspase-3 is required for a-Fodrin cleavage but dispensable for cleavage of other death substrates in apoptosis. J Biol Chem 273:15540–15545

Kawahara A, Enari M, Talanian RV, Wong WW, Nagata S (1998) Fas-induced DNA fragmentation and proteolysis of nuclear proteins. Genes Cells 3:297–306

Keramaris E, Stefanis L, MacLaurin J, Harada N, Kazuaki T, Ishikawa T, Taketo MM, Robertson GS, Nicholson DW, Slack RS, Park DS (2000) Involvement of caspase 3 in apoptotic death of cortical neurons evoked by DNA damage. Mol Cell Neurosci 15:368–379

Kuida K, Zheng TS, Na S, Kuan C-Y, Yang D, Karasuyama H, Rakic P, Flavell RA (1996) Decreased apoptosis in the brain and premature lethality in CPP32-deficient mice. Nature 384:368–372

Kuida K, Haydar TF, Kuan C-Y, Gu Y, Taya C, Karasuyama H, Su MS-S, Rakic P, Flavell RA (1998) Reduced apoptosis and cytochrome c-mediated caspase activation in mice lacking caspase 9. Cell 94:325–337

Kuan CV, Roth KA, Flavell RA, Rakic P (2000) Mechanism of programmed cell death in the developing brain. Trends Neurosci 23:291–297

Li H, Yuan J (1999) Deciphering the pathways of life and death. Curr Opin Cell Biol 11:261–266

Liu P, Nijhawan D, Budihardjo I, Srinivasula SM, Ahmad M, Alnemri ES, Wang X (1997) Cytochrome c and dATP-dependent formation of Apaf-1/caspase-9 complex initiates an apoptotic protease cascade. Cell 91:479–489

Liu X, Li P, Widlak P, Zou H, Luo X, Garrard WT, Wang X (1998) The 40-kDa subunit of DNA fragmentation fractor induces DNA fragmentation and chromatin condensation during apoptosis. Proc Natl Acad Sci USA 95:8461–8466

McCarthy NJ, Whyte MKB, Gilbert CS, Evan GI (1997) Inhibition of Ced-3/ICE-related proteases does not prevent cell death induced by oncogenes, DNA damage, or the Bcl-2 homologue Bak. J Cell Biol 136:215–227

McIlroy D, Tanaka M, Sakahira H, Fukuyama H, Suzuki M, Yamamura K, Ohsawa Y, Uchiyama Y, Nagata S (2000) An auxiliary mode of apoptotic DNA fragmentation provided by phagocytes. Genes Dev 14:549–558

Nagata S (1997) Apoptosis by death factor. Cell 88:355–365

Oppenheim RW (1991) Cell death during development of the nervous system. Ann Rev Neurosci 14:453–501

Oppenheim RW (1999) Programmed cell death. In: Zigmond MJ, Bloom FE, Landis SC, Roberts JL, Squire LR (eds) Fundamental neuroscience. Academic Press, New York, pp 581–610

Oppenheim RW, Flavell RA, Vinsant S, Prevette D, Kuan CY, Rakic P (2001) Programmed cell death of developing mammalian neurons after genetic deletion of caspases. J Neurosci 21:4752–4760

Pilar G, Landmesser L (1976) Ultrastructural differences during embryonic cell death in normal and peripherally deprived ciliary ganglia. J Cell Biol 68:339–356

Roth KA, Yuan CY, Haydar TF, D'Sa-Eipper C, Shindler KS, Zheng TS, Kuida K, Flavell RA, Rakic P (2000) Epistatic and independent functions of caspase-3 and Bcl-X$_L$ in development programmed cell death. Proc Natl Acad Sci USA 97:466–471

Sakahira H, Enari M, Nigata S (1998) Cleavage of CAD inhibitor in CAD activation and DNA degradation during apoptosis. Nature 391:96–99

Shiraiwa N, Shimada T, Nishiyama K, Hong J, Wang S, Momoi T, Uchiyama Y, Oppenheim RW, Yaginuma H (2001) A novel role for caspase-3 activity in early motoneuron death in the click embryo cervical spinal cord. Mol Cell Neurosci (in press)

Stefanis L, Troy CM, Qi H, Shelanski ML, Greene LA (1998) Caspase-2 (Nedd-2) processing and death of trophic factor-deprived PC12 cells and sympathetic neurons occur independently of caspase-3 (CPP32)-like activity. J Neurosci 18(22):9204–9215

Stefanis L, Park SD, Friedman WJ, Greene LA (1999) Caspase-dependent and -independent death of camptothecin-treated embryonic cortical neurons. J Neurosci 19(15):6235–6247

Tanabe K, Nakanish H, Meada H, Nishioku T, Hashimoto K, Liou S-Y, Akamine A, Yamamoto K (1999) A predominant apoptotic death pathway of neuronal PC12 cells induced by activated microglia is displaced by a non-apoptotic death pathway following blockage of caspase-3-dependent cascade. J Biol Chem 274(22):15725–15731

Woo M, Hakem R, Soengas MS, Duncan GS, Shahinian A, Kägi D, Hakem A, McCurrach M, Khoo W, Kaufman SA, Senaldi G, Howard T, Lowe SW, Mak TW (1998) Essential contribution of caspase 3/CPP32 to apoptosis and its associated nuclear changes. Gene Dev 12:806–819

Xue L, Fletcher GC, Tolkovsky AM (1999) Autophagy is activated by apoptosis signaling in sympathetic neurons: an alternative mechanism of death execution. Mol Cell Neurobiol 14:180–198

Yaginuma H, Tomita M, Takashita N, McKay SE, Cardwell C, Yin QW, Oppenheim RW (1996) A novel type of programmed neuronal death in the cervical spinal cord of the chick embryo. J Neurosci 16:3685–3703

Yoshida H, Kong Y-Y, Yoshida R, Elia AJ, Hakem A, Hakem R, Penninger JM, Mak TW (1998) Apaf1 is required for mitochondrial pathways of apoptosis and brain development. Cell 94: 739–750

Zheng TS, Schlosser SF, Dao T, Hingorami R, Crispe IN, Boyer JL, Flavell RA (1998) Caspase-3 controls both cytoplasmic and nuclear events associated with Fas-mediated apoptosis in vivo. Proc Natl Acad Sci USA 95:13618–13623

Caspases and Their Regulation in Apoptosis during Brain Development

C.-Y. Kuan, R. A. Flavell, and P. Rakic

Summary

Cell death is an important mechanism during the mammalian brain development, with documented roles in both morphogenetic and histiogenetic degeneration. The mammalian caspase family of cysteine-containing, aspartate-specific proteases was suggested to play a pivotal role in execution of developmental apoptosis due to its homology to the cell death gene *ced-3* of the nematode *C. elegans*. We have used the gene targeting strategy to test the biological functions of several members of the caspase family. Our results indicate that caspase-9 and caspase-3 are essential for programmed cell death and normal mouse brain development. Moreover, caspase-9 and caspase-3 form a sequential cell death cascade, as the absence of caspase-9 abolishes the cytochrome *c*-mediated caspase-3 activation both in vivo and in vitro and exhibits similar phenotypes to those of caspase-3 deficiency. To test whether the activation of caspase-3 in the mammalian brain is also regulated by Bax and BclxL, homologues of the pro-apoptotic gene *egl-1* and anti-apoptotic gene *ced-9* in *C. elegans*, respectively, we conducted epistatic genetic analysis in *caspase-3/bcl-x* double mutants. The absence of caspase-3 rescued the ectopic cell death of post-mitotic neurons caused by the BclxL deficiency, indicating an evolutionary conserved cell death pathway. However, Bax and BclxL are expressed only in post-mitotic neurons, in contrast to caspase-3, which is activated in the proliferative population and specific brain regions associated with morphogenesis. Moreover, unlike caspase-3 or -9 deficiency, Bax-deficient mutants exhibited a normal amount of cell death in the early developing brain. These results indicate additional signaling pathways in early brain development, preventing a random, haphazard activation of caspases. The Jun N-terminal kinase (JNK) signaling pathway appears to be an important mechanism that regulates the brain region-specific activation of caspases. In the absence of *Jnk1* and *Jnk2* genes, both encoding a somatic form of JNK, there is reduced apoptosis in the hindbrain, leading to neural tube defect and widespread cell death coupled with ectopic caspase-3 activation in the forebrain. Taken together, these results suggest complex regulation mechanisms and distinct functions of caspase activation in mammalian brain development.

Henderson/Green/Mariani/Christen (Eds.)
Neuronal Death by Accident or by Design
© Springer-Verlag Berlin Heidelberg 2001

Introduction

It has long been recognized that cell death plays an important role during normal development of the vertebrate nervous system (Glucksmann 1951; Oppenheim 1991). Traditionally, the investigation of developmental neural death focused on the role of target-derived survival factors such as nerve growth factor (NGF) and related neurotrophins. In the past few years, the genetic analysis of programmed cell death in the nematode *Caenorhabditis elegans* has opened new avenues to understanding the mechanism of neural apoptosis in the mammalian brain (Metzstein et al. 1998). In the core of the programmed cell death pathway in *C. elegans*, *ced-3* plays an essential role in cell killing, the function of which is regulated by *egl-1*, *ced-9* and *ced-4* (Fig. 1). Biochemical studies suggest that Egl-1 triggers programmed cell death by binding to Ced-9 and releases the cell-death activator Ced-4 from a Ced-9/Ced-4 protein complex, subsequently leading to activation of Ced-3 by Ced-4 (Conradt and Horvitz 1998). Remarkably, structural homologues of *egl-1*, *ced-9*, *ced-4* and *ced-3* have all been identified in mammals. The mammalian homologues of *egl-1* and *ced-9* belong to a growing family of Bcl2 proteins that share the Bcl2 homology (BH) domain and are either pro- or anti-apoptotic (Korsmeyer 1999). The *ced-4* homolog is identified as one of the apoptosis-protease activating factors, Apaf (Zou et al. 1997). The mammalian *ced-3* homologs comprise a family of cysteine-containing, aspartate-specific proteases called caspases (Thornberry and Lazebnik 1998). Caspases are expressed in living cells as proenzymes that contain three domains: an NH_2-terminal domain, a large subunit and a small subunit. Activation of caspases involves proteolytic processing between domains followed by association of the large and small subunits to form an active heterodimer or tetra-

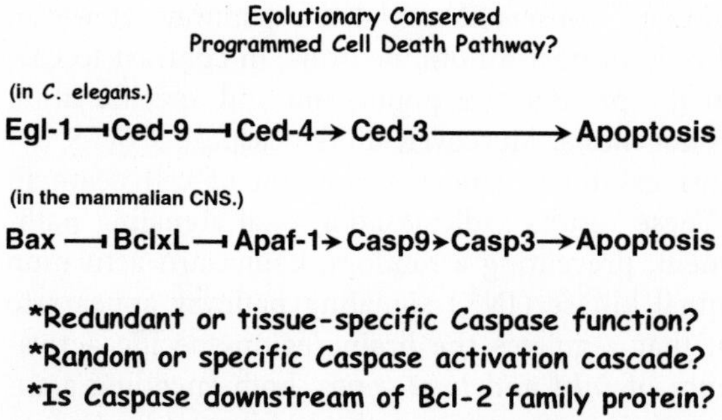

**Evolutionary Conserved
Programmed Cell Death Pathway?**

(in *C. elegans*.)

Egl-1 ⊣ Ced-9 ⊣ Ced-4 → Ced-3 ————→ Apoptosis

(in the mammalian CNS.)

Bax ⊣ BclxL ⊣ Apaf-1 → Casp9 → Casp3 → Apoptosis

***Redundant or tissue-specific Caspase function?
*Random or specific Caspase activation cascade?
*Is Caspase downstream of Bcl-2 family protein?**

Fig. 1. Comparison of the cell-death pathway in the nematode, *Caenorhabditis elegans*, and in mammals. **A** In *C. elegans*, *egl-1*, *ced-9*, *ced-4* and *ced-3* form a linear cascade involved in most, if not all, cells undergoing programmed cell death in the execution phase. **B** In the mammalian central nervous system, Bax, BclxL, Apaf-1, caspase-9 and caspase-3 are identified by gene-targeting studies as the key components of programmed cell death in brain development. However, whether these mammalian homologues constitute an obligate execution cascade for apoptosis in selective or all cell populations is under investigation. (Modified from Kuan et al. 2000)

mer. Once activated, caspases cleave other caspases and various cellular substrates including the DNA fragmentation factor-45/inhibitor of caspase-activated deoxynuclease (DFF45/ICAD), leading to the ultrastructural changes of apoptosis (Liu et al. 1997).

To date, more than 14 members of the caspase family of proteases have been isolated; they have overlapped tissue distribution patterns and share similar cleavage specifity (Thornberry and Lazebnik 1998). The sheer size of this protease family, their overlapping tissue distribution, and a similar cleavage specificity lead to several fundamental questions regarding the biological function of caspases in vivo (Fig. 1). First, is the function of the caspase family members redundant or specific? Second, is there an obligate activation cascade among members of the caspase family? Finally, is the function of caspase downstream to Apaf-1 and Bcl-2 family proteins similar to the relationship of their counterparts in *C. elegans*? In other words, is there an evolutionary conserved cell death pathway in the developing mammalian brain?

Caspase-9 and Caspase-3 Form a Sequential Activation Cascade

Among the several lines of caspase-deficient mutants that have been generated, null mutants of *caspase-3* and *caspase-9* showed the most severe defects of programmed cell death in the nervous system (Kuida et al. 1996, 1998; Hakem et al. 1998). The majority of homozygous *caspase-3* and *-9* null mutants are embryonic lethal or die shortly after birth. A general reduction of pyknotic cell death is found in the embryonic brain of the mutants. As a consequence of the reduction of developmental cell death in the nervous system, multiple indentations of the cerebrum and periventricular masses constituted by supernumerary neurons were generated (Kuida et al. 1996, 1998; Hakem et al. 1998). However, despite the severe defects of programmed cell death in the brain, the developmental apoptosis of thymocytes in the *caspase-3* and *-9* null mutants is largely preserved. Similarly, other lines of caspase-deficient mice (*caspase-1, -2, -8* and *-11*) show preferential apoptosis defects rather than a global suppression of cell death, indicating that individual members of the caspase family have a dominant and non-redundant role in apoptosis in a tissue-selective or stimulus-dependent manner.

The similar phenotypes of null-mutations of *caspase-3* and *-9* suggest that these two caspases might function along the same cell death pathway in brain development. Consistent with this idea, caspase-9 was previously identified as an upstream activator of caspase-3 in a biochemical study using human HeLa cells (Li et al. 1997). Caspase-9 binds to Apaf-1, the human homologue of *ced-4*, and cytochrome *c* through a caspase-recruitment domain (CARD) in its amino-terminal sequence, forming an active apoptosome. By contrast, caspase-3 lacks the CARD motif, does not bind to Apaf-1 directly and is instead activated by the caspase-9/Apaf-1/cytochrome *c* complex (Fig. 2A). Biochemical assay demonstrates that the cytochrome-c-mediated cleavage of pro-caspase-3 is defective in the cytosolic fractions of *caspase-9*

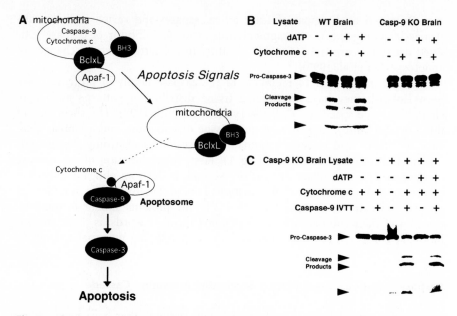

Fig. 2. Activation cascade from caspase-9 to caspase-3 in the developing mouse brain. **A** Schematic diagram of current view of initiation of apoptotic events involving mitochondria. **B** Cytochrome-c-dependent cleavage of pro-caspase-3 cleavage activity was absent in the cytosolic fractions of caspase-9 null mutants. **C** Pro-caspase-3 cleavage activity was restored by the addition of in vitro transcribed-translated (IVTT) caspase-9. KO, Knockout. (Adapted from Kuida et al. 1998)

null mutants but is restored after reconstitution of in vitro transcribed and translated caspase-9 (Fig. 2 B, C). Furthermore, immunofluorescence study reveals an absence of the active form of caspase-3 in the *caspase-9*-deficient embryonic brain (Kuida et al. 1998). Taken together, these results indicate a linear activation cascade from caspase-9 to caspase-3 in developmental programmed cell death in the mammalian brain.

Caspase-3 and BclxL have Both Independent and Epistatic Apoptotic Functions

In addition to *caspase-3* and *caspase-9* null mutations, gene targeting studies indicate that Bax, BclxL and Apaf-1 deficiencies also cause abnormality of programmed cell death in the mammalian nervous system. Targeted disruption of *bax* dramatically decreases programmed cell death in the developing nervous system, which results in increased numbers of neurons in selected neuronal populations. Bax-deficient neurons showed decreased susceptibility to trophic-factor withdrawal both in vivo and in vitro (Knudson et al. 1995; Deckwerth et al. 1996). Targeted disruption of *bcl-x* caused a dramatic increase in apoptosis of immature neurons throughout the embryonic nervous system but failed to affect apoptosis of neural precursor cells in the ventricular zone. In addition, BclxL-deficient embryos die at approximately E 13.5, secondary to increased hematopoietic apoptosis (Motoyama et al. 1995). Insertional mutation or targeted disruption of *apaf-1* produced perinatal lethality that was associated with marked craniofacial abnormalities, alterations in the eye and retina, and a variety of brain abnormalities, including exencephaly, hyperplasia and ectopic neural masses (Cecconi et al. 1998; Yoshida et al. 1998). Apaf-1 deficient cells fail to activate caspase-3 in vivo or in vitro, which presumably accounts for the similar developmental abnormalities seen in Apaf-1, caspase-9- and caspase-3-deficient embryos (Cecconi et al. 1998).

The above-mentioned gene-targeting studies thus identify Bax, BclxL, Apaf-1, caspase-9 and caspase-3 as key regulators of programmed cell death in neural development. Moreover, on the basis of the sequence homology and analogous functions, Bax, BclxL, Apaf-1, caspase-9 and caspase-3 may form an evolutionarily conserved cell death pathway in the mammalian nervous system (Fig. 1). To test this hypothesis, we conducted epistatic genetic analysis in *caspase-3* and *bcl-x* double mutants to examine whether the concomitant caspase-3 deficiency would eliminate the increased ectopic cell death caused by the BclxL deficiency as predicted by an epistatic relationship of these two molecules (Roth et al. 2000).

In normal development, clusters of pyknotic cells are confined to specific locations of the nervous system at precise times of development (Fig. 3 A). This brain region-specific apoptosis presumably affects neuronal progenitor cells, given its occurrence during the period of active neurogenesis. Except for these brain region-specific cell deaths, only a few pyknotic cells are sparsely distributed in the rest of the developing nervous system (Fig. 3 B,C). By contrast, *bcl-x* deficiency causes widespread pyknotic clusters in the postmitotic population in addition to the region-specific cell death in the developing brain (Fig. 3 F). When *bcl-x* mutant mice are crossed to the *caspase-3* deficiency genetic background, the double mutation of *bcl-x* and *caspase-3* virtually abrogates the ectopic cell death caused by the *bcl-x* deficiency alone (Fig. 3 I). Similarly, the concomitant *caspase-3* deficiency prevents the increased apoptosis of *bcl-x*-deficient cortical neurons in response to serum deprivation in vitro (Roth et al. 2000). These results suggest that *bcl-x* deficiency causes apoptosis of postmitotic neurons primarily through uninhibited activation of caspase-3. However, the epistatic relationship is clearly not universal, as *caspase-3* deficiency does not prevent increased hematopoietic cell apoptosis and embryonic lethality in *bcl-x*-deficient mice, suggesting a caspase-3-independent apoptotic pathway that is normally suppressed by BclxL during development of the hematopoietic system.

If Bax and caspase-3 are both pro-apoptotic in an obligate, epistatic cell death pathway, the phenotype of *bax* deficiency should be identical to that of *caspase-3* deficiency. On the contrary, there are significant differences between the phenotypes of *bax*- and *caspase-3*-deficient embryos. There are no signs of hyperplasia or malformations of the nervous system, and the apoptosis of neuronal progenitor cells, which is exemplified by brain region-spe-

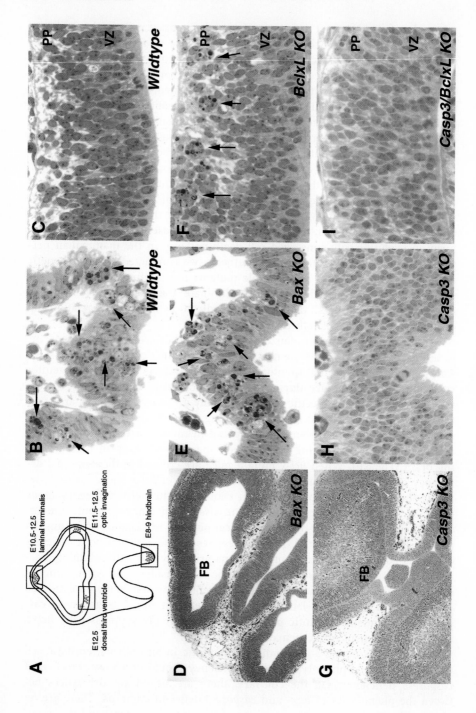

cific apoptosis, is preserved in the *bax*-deficient embryos (Fig. 3 D, E). By contrast, the *caspase-3* deficiency greatly reduces the brain region-specific apoptosis and shows marked hyperplasia of the embryonic nervous tissue (Fig. 3 G, H). These results indicate, therefore, that, though downstream of Bax and BclxL in the apoptosis of postmitotic neurons, caspase-3 has a unique function in regulating the size of the progenitor pool during early development, even before neurogenesis in a given region begins.

Founder Cell Versus Post-mitotic Cell Apoptosis

On the basis of the phenotype of individual mutants and epistatic genetic analysis, a scheme is proposed to explain the interactions of Bax, BclxL and caspase-3 during mammalian brain development (Fig. 4). BclxL inhibits the pro-apoptotic effect of caspase-3 in the postmitotic neuronal population, and therefore *bcl-x* deficiency causes increased apoptosis of postmitotic neurons, which is prevented by the additional absence of *caspase-3*. Bax modulates the anti-apoptotic effects of BclxL; the null-mutation of *bax* therefore reduces the normally occurring developmental death of postmitotic neurons without affecting the global formation of the nervous system. The unique feature of caspase-3 in this scheme is its dual function in both postmitotic and neuronal progenitor apoptosis. Although *caspase-3* deficiency results in decreased apoptosis of postmitotic neurons in the developing cortex, given the normal brain organization in *bax*-deficient mice, this effect is insufficient to cause gross malformations. Rather, caspase-3 deficiency rescues a number of the progenitor cells from programmed cell death, which results in an exponential expansion of the progeny, ultimately leading to marked dysplasia and malformations of the nervous system.

How early does programmed cell death occur in the embryonic nervous tissue? As early as E 8.5 day in mouse embryo when the neural tube is preceding the "turning" sequence, there are substantial numbers of pyknotic nuclei in the neuroepithelial plate (Fig. 5). The process of "turning" is usually initiated when the mouse embryo possesses six to eight pairs of somites and

Fig. 3. Comparison of the developmental brain apoptosis in wild-type, *bax*-deficient, *bcl-x*-deficient, *caspase-3*-deficient and *bcl-x/caspase-3* double mutant embryos. In normal embryogenesis, intense apoptosis occurs in a spatial and temporal-precise manner. **A** For example, clusters of pyknotic cells (indicated by arrows in **B** are found in wild-type embryos in the laminal terminals **B** but not in either the postmitotic preplate (PP) or the proliferative ventricular zone (VZ) of the developing cortical wall **C**. Mice deficient in *bax* exhibit an apparently normal forebrain (FB) formation **D** and retain numerous pyknotic cells in the laminal terminalis **E** at embryonic (E) day 12. The bcl-x deficiency causes ectopic death of postmitotic neurons in the preplate but not in the proliferative ventricular zone in E12.5 developing cortex **F**. In contrast to the phenotype of *bax* deficiency, mice deficient in *caspase-3* show severe hyperplasia of the embryonic forebrain **G** and absence of pyknotic cells in the laminal terminalis **H**. Moreover, the ectopic pyknotic cell death in the developing cortex caused by *bcl-x* null-mutation is rescued by the *bcl-x/caspase-3* double deficiency **I**. (Reproduced from Kuan et al. 2000)

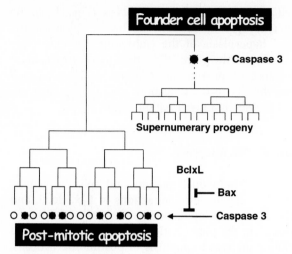

Fig. 4. Distinct molecular mechanism of apoptosis of the neuronal founder cells and postmitotic neurons in the developing mammalian nervous system. Gene-targeting studies and epistatic genetic analysis indicate that Bax, BclxL and caspase-3 form an obligate cell-death pathway in the postmitotic neurons. By contrast, neither Bax nor BclxL is involved in the apoptosis of neuronal founder cells, which is greatly reduced by the caspase-3 deficiency, resulting in the generation of supernumerary progeny and severe brain malformations as a consequence. (Reproduced from Kuan et al. 2000)

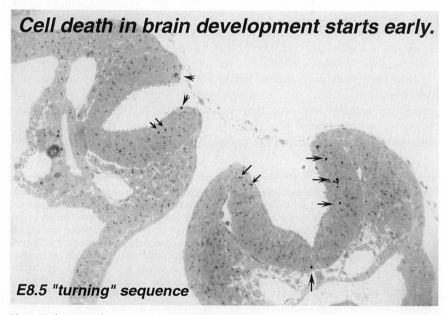

Fig. 5. Early onset of programmed cell death in the developing mammalian nervous system. Pyknotic cell deaths are readily detectable (indicated by arrows) in the proliferative neuroepithelium when the neural tube undergoes the "turning" sequence (or axial rotation)

is normally completed by the time that the embryo possesses 14–16 pairs of somites (Kaufman 1992). These early deaths within the proliferative neuro-epithelial plate would thus influence the morphogenetic process of the mammalian nervous system. Moreover, the identification of cell death as a regulatory mechanism of founder cell numbers in the neural tube has important implications for the development and evolution of the mammalian forebrain. The telencephalic expansion during mammalian evolution is resulted from massive enlargement and complex morphogenesis of the forebrain portion of the neural tube, which is visible even before the onset of postmitotic neuron generation. Although neurogenesis is the main engine for this telencephalic expansion, the increased forebrain size in caspase-3- and -9-deficient embryos suggests that programmed cell death might participate in determining the production of certain progenitor populations while sparing others. Therefore, the precise coordination of proliferation and differential apoptosis mediated by caspases during early neurogenesis is crucial for the proper regulation of cortical size and shape in the mammalian brain (Haydar et al. 1999; Rakic 1995).

Apoptosis Deficiency Causes Neural Tube Defects

One important feature of programmed cell death is its occurrence at precise times and places during normal brain development (Kallen 1955). Intriguingly, although inactive proenzymes of caspases are widely expressed in the nervous system, intense caspase activation only occurs at restricted locations. Thus, there exists either a neuroprotective mechanism to prevent excessive caspase activation and/or specific proapoptotic induction at restricted locations. The temporal and spatial precision of cell death along the neuraxis is needed for the proper pruning of embryonic tissues (morphogenetic degeneration; Glucksmann 1951).

Mounting evidence indicates that this morphogenetic degeneration may be essential for hindbrain neural tube closure. First, the blocking of apoptosis by caspase inhibitors causes failure of neural tube closure in cultured chick embryos (Weil et al. 1997). Secondly, a portion of caspase-3-deficient embryos exhibits prominent neural tube defects, including hindbrain exencephaly and spinal bifida (aperta) in the caudal spinal cord (Fig. 6). Hindbrain exencephaly defect is also found in *caspase-9, apaf-1* null-mutants (Kuida et al. 1998; Cecconi et al. 1998; Hakem et al. 1998; Yoshida et al. 1998) and mice lacking two somatic forms of c-Jun N-terminal kinase genes *Jnk1* and *Jnk2* (Kuan et al. 1999; Sabapath et al. 1999). In the case of *Jnk1/Jnk2* double mutants, a reduction of apoptosis in the hindbrain has been demonstrated and attributed to the cause of exencephaly abnormality.

The Jnk family of protein kinases was first identified as a subfamily of the MAP kinases. Jnk mediates the phosphorylation of c-Jun at serines 63 and 73 and causes increased AP-1 transcription activity in response to external stress signals (for review, see Ip and Davis 1998). The Jnk family has

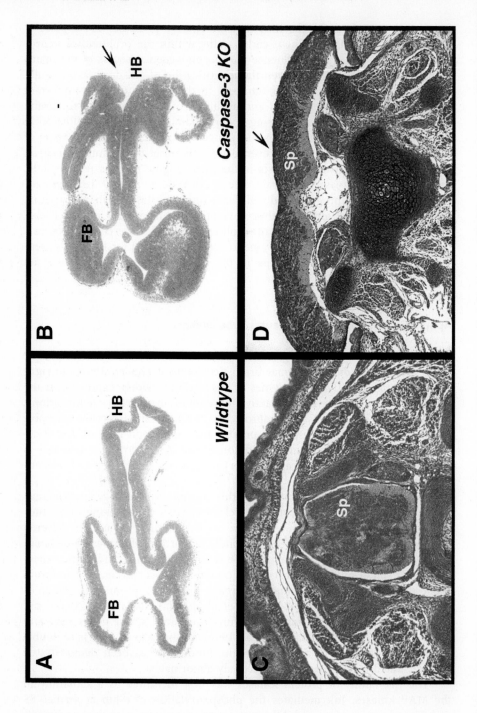

three different isoforms (*Jnk1, Jnk2,* and *Jnk3*) that have distinct substrate affinities and tissue distributions (Gupta et al. 1996). Jnk1 and Jnk2 are widely expressed in adult tissues whereas Jnk3 is mainly enriched in the nervous system (Martin et al. 1996). Although *Jnk1, Jnk2,* or *Jnk3* single deficient and *Jnk1/Jnk3* or *Jnk2/Jnk3* double mutants all survived normally, *Jnk1/Jnk2* dual-deficient mutants died on E11–12 (Kuan et al. 1999). Histological examination of these mutants revealed a great reduction in pyknotic cells at the lateral edges of the folding hindbrain prior to neural tube closure at E9 (Fig. 7A,B). Mechanistically, the focal apoptosis at the lateral edges of the closing neural tube may reduce the number of progenitor cells and subsequently create a mediolateral imbalance of proliferation, leading to apposition of bilateral neural folds. Presumably as a consequence of the reduction of cell death at the lateral edge of neural plate, the mutant embryos exhibit hindbrain exencephaly at E11.5 (Fig. 7C,D).

Two additional global morphogenetic processes occurring at the same time, namely the downward cephalic flexure and the axial rotation of the fetal body, may also facilitate the closure of the hindbrain (Fig. 7E). During cranial neurulation, the neural tube is first closed at the border between spinal cord and hindbrain and in the forebrain (de novo closure) followed by a progressive apposition of the hindbrain neural tube in both forward and backward directions (continuation closure). The concomitant rotations of the body axis may generate a mechanical tension on the hindbrain and contribute to its closure when there is less and weaker cell mass at the edges. By contrast, when there is reduction of cell death at the lateral neural folds, the mechanic tension generated by axial rotation may bend the initially V-shaped hindbrain neural plate into the final biconvex configuration as seen in the *Jnk1/Jnk2* double mutants (Fig. 7E). It should nevertheless be emphasized that neural tube closure is a complex morphogenetic event requiring a multitude of factors in addition to programmed cell death (Copp et al. 1990; Smith and Schoenwolf 1997; Harris and Juriloff 1999).

Concluding Remarks

Genetic studies of programmed cell death in the nematode *C. elegans* provided new approaches to study the mechanism of apoptosis in mammalian neural development. A cumulative body of evidence indicates that mammalian homologues of the cell death genes in *C. elegans* have analogous functions in apoptosis and form a similar epistatic pathway in brain development. Moreover, these studies also implicate caspase-3, and presumably cas-

Fig. 6. Neural tube defects are part of the phenotypes of *caspase-3* deficiency. A portion of caspase-3-deficient embryos exhibit hindbrain exencephaly at E11.5 day **B** and spinal bifida (aperta) abnormality in the caudal spinal cord at E14.5 **D**, standing in contrast to the normal neural tube closure of wildtype embryos of the same age **A, C**

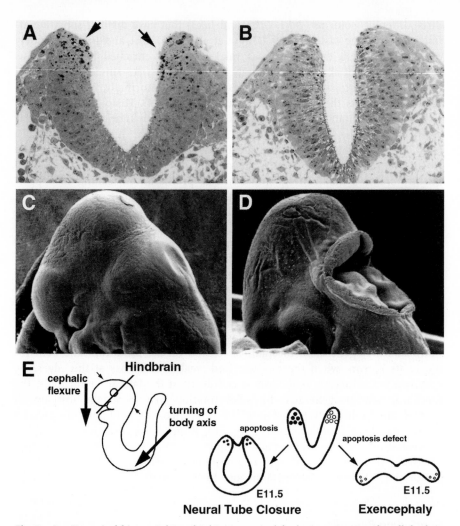

Fig. 7. c-Jun N-terminal kinases Jnk1 and Jnk2 are required for brain region-specific cell death in the hindbrain during neural tube closure. **A,B** Pyknotic cell deaths are typically located at the lateral edges of the hindbrain prior to neural tube closure in wild type embryo **A** but are greatly reduced in mouse mutants deficient in both Jnk1 and Jnk2 genes **B. C,D** As a consequence of this reduction in region-specific apoptosis, Jnk1 and Jnk2 dual-deficient embryos exhibited neural tube defects at the hindbrain. **E** Two global morphogenetic movements, cephalic flexure and rotation of body axis, create a bending tension that either propes the closure of hindbrain neural tube or leads to exencephaly, depending in part on whether cell death occurs on the lateral edges of the folding neural plate. (Modified from Kuan et al. 1999)

pase-9 and Apaf-1 as well, in the apoptosis of neuronal progenitor cells, a function that is distinct form the classic role of programmed cell death in matching postmitotic neuronal population with post-synaptic targets. These findings thus provide new insights to the biological functions of programmed cell death in brain development and raise some intriguing questions. As the Bcl2-family proteins Bax and BclxL are not involved in the caspase-3-mediated early progenitor cell death, how is the early brain region-specific apoptosis regulated so precisely, especially given the ubiquitous presence of caspases throughout the nervous system? In theory, there could be novel signal-transduction mechanisms to trigger apoptosis at specific locations or general cytoprotective mechanism to prevent excessive activation of caspases in the rest of the nervous system. The identification of the cytoprotective mechanism and brain region-specific apoptotic signaling inductions raises new challenges in developmental biology.

Acknowledgments. The authors thank their colleagues, especially K. Kuida, D. D. Yang, T. F. Haydar, K. A. Roth, R. J. Davis and M. S.-S. Su, for many stimulating discussions leading to the ideas presented in this article. The research in the laboratories of P.R. and R.A.F. is supported by grants from the Public Health Service.

References

Cecconi F, Alvarez-Bolado G, Meyer BI, Roth KA, Gruss P (1998) Apaf1 (CED-4 Homolog) regulates programmed cell death in mammalian development. Cell 94:727–737

Conradt B, Horvitz HR (1998) The *C. elegans* protein EGL-1 is required for programmed cell death and interacts with the Bcl-2-like protein CED-9. Cell 93:519–529

Copp AJ, Brook FA, Estibeiro JP, Shum AS, Cockroft DL (1990) The embryonic development of mammalian neural tube defects. Prog Neurobiol 35:363–403

Deckwerth TL, Elliott JL, Knudson CM, Johnson EM Jr, Snider WD, Korsmeyer SJ (1996) Bax is required for neuronal death after trophic factordeprivation and during development. Neuron 17:401–411

Glucksmann A (1951) Cell deaths in normal vertebrate ontogeny. Bio Rev 26:59–86

Gupta S, Barrett T, Whitmarsh AJ, Cavanagh J, Sluss HK, Derijard B, Davis RJ (1996) Selective interaction of JNK protein kinase isoforms with transcription factors. EMBO J 15:2760–2770

Hakem R, Hakem A, Duncan GS, Henderson JT, Woo M, Soengas MS, Elia A, de la Pompa JL, Kagi D, Khoo W, Potter J, Yoshida R, Kaufman SA, Lowe SW, Penninger JM, Mak TW (1998) Differential requirement for caspase-9 in apoptotic pathways in vivo. Cell 94:339–352

Harris MJ, Juriloff DM (1999) Mini-review: toward understanding mechanisms of genetic neural tube defects in mice. Teratology 60:292–305

Haydar TF, Kuan CY, Flavell RA, Rakic P (1999) The role of cell death in regulating the size and shape of the mammalian forebrain. Cereb Cortex 9:621–626

Ip YT, Davis RJ (1998) Signal transduction by the c-Jun N-terminal kinase (JNK)-from inflammation to development. Curr Opin Cell Biol 10:205–219

Kallen B (1955) Cell degeneration during normal ontogenesis of the rabbit brain. J Anat 89:153–161

Kaufman MH (1992) The atlas of mouse development. Academic Press, San Diego

Knudson CM, Tung KS, Tourtellotte WG, Brown GA, Korsmeyer SJ (1995) Bax-deficient mice with lymphoid hyperplasia and male germ cell death. Science 270:96–99

Korsmeyer SJ (1999) BCL-2 gene family and the regulation of programmed cell death. Cancer Res 59:1693s–1700s

Kuan CY, Yang DD, Samanta Roy DR, Davis RJ, Rakic P, Flavell RA (1999) The Jnk1 and Jnk2 protein kinases are required for regional specific apoptosis during early brain development. Neuron 22:667–676

Kuan CY, Roth KA, Flavell RA, Rakic P (2000) Mechanisms of programmed cell death in the developing brain. Tends Neurosci 23:291–297

Kuida K, Zheng TS, Na S, Kuan C, Yang D, Karasuyama H, Rakic P, Flavell RA (1996) Decreased apoptosis in the brain and premature lethality in CPP32-deficient mice. Nature 384:368–372

Kuida K, Haydar TF, Kuan CY, Gu Y, Taya C, Karasuyama H, Su MS, Rakic P, Flavell RA (1998) Reduced apoptosis and cytochrome c-mediated caspase activation in mice lacking caspase-9. Cell 94:325–337

Li P, Nijhawan D, Budihardjo I, Srinivasula SM, Ahmad M, Alnemri ES, Wang X (1997) Cytochrome c and dATP-dependent formation of Apaf-1/Caspase-9 complex initiates an apoptotic protease cascade. Cell 91:479–489

Liu X, Zou H, Slaughter C, Wang X (1997) DFF, a heterodimeric protein that functions downstream of caspase-3 to trigger DNA fragmentation during apoptosis. Cell 89:175–184

Martin JH, Mohit AA, Miller CA (1996) Developmental expression in the mouse nervous system of the p493F12 SAP kinase. Brain Res Mol Brain Res 35:47–57

Metzstein MM, Stanfield GM, Horvitz HR (1998) Genetics of programmed cell death in C. elegans: past, present and future. Trends Genet 14:410–414

Motoyama N, Wang F, Roth KA, Sawa H, Nakayama K, Negishi I, Senju S, Zhang Q, Fujii S, Loh DY (1995) Massive cell death of immature hematopoietic cells and neurons in Bcl-x-deficient mice. Science 267:1506–1510

Oppenheim RW (1991) Cell death during development of the nervous system. Annu Rev Neurosci 14:453–501

Rakic P (1995) A small step for the cell, a giant leap for mankind: a hypothesis of neocortical expansion during evolution. Trends Neurosci 18:383–388

Roth KA, Kuan C, Haydar TF, D'Sa-Eipper C, Shindler KS, Zheng TS, Kuida K, Flavell RA, Rakic P (2000) Epistatic and independent functions of Caspase-3 and Bcl-X$_L$ in developmental programmed cell death. Proc Natl Acad Sci USA 97:466–471

Sabapathy K, Jochum W, Hochedlinger K, Chang L, Karin M, Wagner EF (1999) Defective neural tube morphogenesis and altered apoptosis in the absence of both JNK1 and JNK2. Mech Dev 89:115–124

Smith JL, Schoenwolf GC (1997) Neurulation: coming to closure. Trends Neurosci 20:510–517

Thornberry NA, Lazebnik Y (1998) Caspases: enemies within. Science 281:1312–1316

Weil M, Jacobson MD, Raff MC (1997) Is programmed cell death required for neural tube closure? Curr Biol 7:281–284

Yoshida H, Kong YY, Yoshida R, Elia AJ, Hakem A, Hakem R, Penninger JM, Mak TW (1998) Apaf-1 is required for mitochondrial pathways of apoptosis and brain development. Cell 94:739–750

Zou H, Henzel WJ, Liu X, Lutschg A, Wang X (1997) Apaf-1, a human protein homologous to C. elegans CED-4, participates in cytochrome c-dependent activation of caspase-3. Cell 90:405–413

Apoptotic Neurodegeneration in the Developing Brain

C. Ikonomidou

Summary

Apoptosis is a mechanism of crucial importance for normal brain development. In the immature mammalian brain and during a period of rapid growth (brain growth spurt/synaptogenesis period), neuronal apoptosis can be triggered by trauma or interference in the action of neurotransmitters. Transient blockade of glutamate N-methyl-D-aspartate (NMDA) receptors, or activation of gamma-aminobutyric acid (GABA$_A$) receptors, causes millions of developing neurons to commit suicide. Agents that can trigger apoptotic neurodegeneration in the developing mammalian brain include anesthetics (ketamine, nitrous oxide, halothane, isoflurane, propofol), anticonvulsants (benzodiazepines, barbiturates) and drugs of abuse (ethanol, phencyclidine, ketamine). Ethanol, the most widely abused drug in the world, is both an NMDA antagonist and a GABA$_A$ agonist and can therefore trigger massive apoptotic neurodegeneration in the developing brain. This mechanism can explain the reduced brain mass and lifelong neurobehavioral disturbances associated with intrauterine exposure of humans to ethanol (fetal alcohol syndrome). In addition, since many of these apoptogenic agents are used in pediatric and obstetrical medicine for purposes of sedation, anesthesia and treatment of seizure disorders, further research will be required to ascertain the risk posed by exposing the developing human brain to compounds that act by this mechanism during the brain growth spurt period, which in humans starts in the third trimester of pregnancy and extends to several years after birth.

Introduction

Damage to the brain during critical stages of development is thought to constitute one important component in the pathogenesis of neuropsychiatric and neuroendocrine disturbances. Excitotoxic neurodegeneration in response to exogenous toxins represents one such mechanism. Three decades ago it was discovered (Olney 1969) that oral or subcutaneous administration of glutamate to infant animals at doses that do not disrupt normal behaviors quietly destroyed many neurons in the developing hypothalamus. In adolescence, these animals displayed various disturbances in neuroendocrine function, in-

Henderson/Green/Mariani/Christen (Eds.)
Neuronal Death by Accident or by Design
© Springer-Verlag Berlin Heidelberg 2001

cluding hypogonadism, infertility and a reduced size of the pituitary gland. These abnormalities can be explained by the fact that the hypothalamic neurons deleted during infancy assume important neuroendocrine functions that mature and become functional in adolescence and early adulthood.

A newly discovered mechanism that is able to inflict silent damage to the developing brain during the critical stage of synaptogenesis is apoptotic neurodegeneration, which can be triggered either by brain insults (trauma, hypoxia/ischemia) or by interference with normal neurotransmitter function.

Physiological Apoptosis in the Mammalian Brain

Apoptotic cell death, as it was originally described by Kerr and colleagues (1972), displays very distinct morphological features. In the presence of an intact cytoplasmic membrane, the nuclear chromatin starts to form clumps that attach to the nuclear membrane, the cell progressively condenses and it finally breaks down into fragments (apoptotic bodies) that are phagocytosed by neighboring cells in the absence of an inflammatory reaction.

The prototypic example of apoptosis in the mammalian brain is physiological (programmed) cell death, which represents the natural process by which biologically redundant neurons are deleted during development.

Ishimaru et al. (1999) examined the ultrastructural features of neurons undergoing programmed cell death in the in vivo mammalian brain and defined four stages. This study and studies from other laboratories (Gwag et al. 1997; Charriaut-Marlangue and Ben-Ari 1995; Grasl-Kraupp et al. 1995; Collins et al. 1992) showed that a number of methods, i.e., TUNEL staining and DNA electrophoresis, cannot distinguish between excitotoxic and apoptotic neurodegeneration. The conclusion was drawn that detailed analysis of the ultrastructural features of degenerating cells by electron microscopy is required to define a given neurodegenerative process as apoptotic. This is the approach we have been taking (Ishimaru et al. 1999). Using this approach and taking as reference standards the ultrastructural criteria for neuronal apoptosis described by Ishimaru et al., we found that, during the synaptogenesis (brain growth spurt) period, developing neurons are exceedingly prone to delete themselves by apoptosis in response to a variety of insults.

Trauma and NMDA Antagonists Induce Apoptotic Neurodegeneration in the Developing Brain

To study aspects of traumatic injury to the immature brain we developed a model of concussive head trauma in infant rats (Ikonomidou et al. 1996; Pohl et al. 1999). We found that a concussive force applied to the skull overlying the parietal cortex of seven-day-old rats caused two types of lesions: a relatively small excitotoxic lesion at the impact site and disseminated apopto-

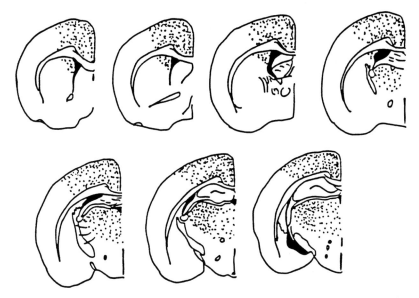

Fig. 1. Distribution pattern of apoptotic neurodegeneration 24 hrs following focal trauma to the right parietal cortex in seven-day-old rats in the hemisphere ipsilateral to the trauma

tic lesions at distant sites. The excitotoxic lesion evolved very rapidly (within 4 hrs after impact) to end-stage neuronal necrosis and remained localized to the site of primary impact. Administration of the NMDA antagonist MK801 prior to concussive injury protected against this acute excitotoxic lesion at the impact site (Ikonomidou et al. 1996). Over the ensuing 24 hrs additional foci of neurodegeneration developed at distant sites, mainly ipsilateral to the impact (Fig. 1). This delayed neurodegenerative response was of higher magnitude than the excitotoxic response (100-fold greater) and subsided by 48 hrs after the traumatic insult. Ultrastructural analysis indicated that the degenerating neurons at these sites qualified for a diagnosis of apoptosis (Pohl et al. 1999). In an attempt to prevent delayed apoptotic neurodegeneration following trauma using various neuroprotective strategies, we were surprised to find that NMDA antagonists not only failed to protect against the apoptotic response, they substantially increased the magnitude of this response (Pohl et al. 1999).

This finding prompted the question of whether NMDA antagonists might also promote the spontaneous apoptotic neurodegenerative process that occurs naturally (independent of head trauma) in the normal developing brain. We investigated this possibility and found that MK801, when administered to 7-day-old infant rats, triggers a massive apoptotic neurodegenerative response affecting many neurons in several major regions of the developing brain (Ikonomidou et al. 1999). Similarly, administration of the NMDA antagonists phencyclidine and ketamine to seven-day-old infant rats by dosing regimes calculated to keep the rat pups intoxicated for up to 8 hrs triggered

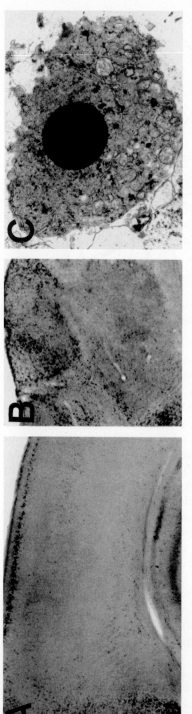

Fig. 2. Panels **A** and **B** are low magnification (×25) light microscopic overviews of silver-stained transverse sections from the parietal and cingulate cortex (**A**) and thalamus (**B**) of eight-day-old rats treated 24 hrs previously with diazepam. Degenerating neurons (small dark dots) are abundantly present in several brain regions following diazepam treatment. Panel C is an electron micrograph (×1800) illustrating that neurodegeneration triggered by diazepam has features consistent with apoptosis

a robust neurodegenerative response in the developing brain (Ikonomidou et al. 1999).

In additional experiments, it was determined that the time window of vulnerability to the apoptosis-inducing action of NMDA antagonists coincides with the period of synaptogenesis, also known as the brain growth spurt period. This period in the rat is largely confined to the postnatal period; it begins one day before birth and terminates at 14 days after birth. The comparable period in humans spans the last three months of pregnancy and extends into the first several years after birth (Dobbing and Sands 1979). Within the brain growth spurt period different neuronal populations become sensitive at different times to the mechanism by which NMDA antagonists trigger apoptotic degeneration (Ikonomidou et al. 1999). Depending on whether exposure occurs in the early, mid or late stage of the brain growth spurt period, different combinations of neuronal groups will be deleted from the brain. Thus, this neurodevelopmental mechanism has the potential to produce a great variety of neurobehavioral deficit syndromes.

Ethanol and GABAergic Agents Induce Apoptotic Neurodegeneration in the Developing Brain

Ethanol, the most widely abused drug in the world, has been known for decades to damage the developing human fetus. Despite extensive research, the pathophysiology of ethanol's neurotoxic effects in the immature brain remained unclear. Evidence that ethanol has NMDA antagonist properties (Lovinger and White 1991; Hoffman et al. 1989; White et al. 1990; Chandler et al. 1994) prompted us to evaluate its ability to mimic the proapoptotic effects of other NMDA antagonists. Administration of ethanol to seven-day-old infant rats triggered a neurodegenerative response that is even more robust than the response to MK801 (Olney et al. 1998; Ikonomidou et al. 2000). Evaluation of the ethanol-induced degenerative response by electron microscopy revealed that it conforms to the criteria for apoptotic cell death. The window of vulnerability to ethanol-induced apoptosis was found to be the same as that for MK801 (coincides with the synaptogenesis/brain growth spurt period). In addition, within the brain growth spurt period different neuronal populations become sensitive at different times to the mechanims by which ethanol triggers apoptotic degeneration. It was also determined that maintaining blood ethanol concentrations at or above 200 mg/dl for four consecutive hours was the minimum condition for triggering neurodegeneration, and if ethanol concentrations remained above 200 mg/dl for more than four hours the degenerative response became progressively more severe and more widespread in proportion to how long the concentrations remained above this level.

Because ethanol triggered apoptosis in some brain regions not typically affected by NMDA antagonists, we attempted to identify other possible mechanisms to explain ethanol's effects in these brain regions. We were unable to demonstrate an appreciable apoptotic response to agents that act as either agonists or antagonists at dopamine receptors, or that block kainic acid or muscarinic cholinergic receptors or that block voltage gated ion channels, but a robust apoptotic response was triggered by agents (benzodiazepines and barbiturates) that mimic or potentiate the action of GABA at $GABA_A$ receptors. The agents tested were diazepam, clonazepam, pentobarbital, and phenobarbital. These agents, in a dose-dependent manner, triggered widespread apoptotic cell death in the infant rat brain. The distribution pattern of degeneration was similar for each GABAergic agent and differed in several major respects from that induced by NMDA antagonists. Superimposing one pattern upon the other resulted in a composite pattern closely resembling that induced by ethanol (Ikonomidou et al. 2000).

Significance of these Findings in a Public Health and Medical Context

Ethanol is, and has been for thousands of years, the most widely abused drug in the world. It has caused more neurodevelopmental morbidity in human offspring than any other agent in the human environment. The fetal alcohol syndrome (FAS) was described in the 1970s and is characterized by craniofacial anomalies, microcephaly and mental retardation. FAS constitutes the more severe form of impairment due to intrauterine exposure to ethanol, whereas less severe cases are referred to as fetal alcohol effects (FAE; Jones and Smith 1973; Jones et al. 1973; Clarren et al. 1978; Sulik et al. 1981; Kerns et al. 1977; Eckardt et al. 1998).

The discovery that transient exposure of the in vivo mammalian brain to ethanol during the synaptogenesis period causes tens of millions of neurons to commit suicide provides an explanation for the reduced brain mass and neurobehavioral disturbances associated with human FAE/S, in that the human synaptogenesis/brain growth spurt period includes the last three months of gestation (Dobbing and Sands 1979). Furthermore, the blood ethanol levels required to trigger apoptotic neurodegeneration in the immature rat brain (200 mg/dl lasting four or more hours) are in the range that a third trimester human fetus might be exposed to by a pregnant mother who consumes ethanolic beverages for several hours in a single drinking episode. Hyperactivity/attention deficit disorder, learning disabilities and mental retardation are neurobehavioral disturbances associated with FAS/FAE. However, Famy et al. (1998) recently studied adult subjects who received a FAE/S diagnosis in childhood and found that a high percentage (72%) required psychiatric attention as adults for a variety of adult-onset psychiatric problems, including a 40% incidence of psychosis and 44% incidence of major depressive disorder. Thus, ethanol is a prime example of an agent that can quietly delete large numbers of neurons from the developing brain and give rise to neurobehavioral and psychiatric disturbances that will partly manifest in adolescence and adulthood. Within the brain growth spurt period different neuronal populations have different temporal patterns for responding to the apoptosis-inducing effects of ethanol. Thus, depending on the timing of exposure, different combinations of neuronal groups will be deleted, which explains why fetal ethanol exposure gives rise to a wide spectrum of neuropsychiatric disturbances (Famy et al. 1998).

Apart from being drugs of abuse, NMDA antagonists and GABA$_A$ agonists are also sedatives, anesthetics, tranquilizers and anticonvulsants used in obstetric and pediatric medicine (Jevtovic-Todorovic et al. 1998). In recent work we performed to determine apoptogenic threshold doses of barbiturates and benzodiazepines we found that doses within the anticonvulsant range are capable of triggering an apoptotic response in the developing rat brain (Bittigau et al. 2000). Thus, it appears likely that long-term treatment of human epilepsies with barbiturates and benzodiazepines during late pregnancy or in the first years of life bears the danger of deleting immature neurons by apoptosis. This mechanism can explain reduced head circumference and cognitive impairment in children of women with epilepsy and in humans who received treatment with phenobarbital during their first years of life (Zahn et al. 1998; Dessens et al. 2000).

In the context of pediatric anesthesia, combinations of drugs with NMDA antagonistic (ketamine, nitrous oxide) and GABA$_A$ agonistic properties (propofol, isoflurane, benzodiazepines, barbiturates) are used at doses that impair conciousness and thus mimic an intoxicated state caused by ethanol. Further studies will be necessary to determine to what extent established anesthetic practices impose risks to the developing human brain in that they can silently delete large numbers of neurons.

References

Bittigau P, Genz K, v Engelbrechten-Ilow S, Hoerster F, Dikranian K, Tenkova T, Olney JW, Ikonomidou C (2000) Antiepileptics which enhance GABAergic inhibition cause neuronal apoptosis in the developing CNS. Soc Neurosci Abst 26:323

Chandler LJ, Guzman NJ, Sumers C, Crews FT (1994) Magnesium and zinc potentiate ethanol inhibition of NMDA-stimulated nitric oxide synthase in cortical neurons J Pharm Exp Ther 271:67–75

Charriaut-Marlangue C, Ben-Ari Y (1995) A cautionary note on the use of the TUNEL stain to determine apoptosis. Neuroreport 7:61–64

Clarren SK, Alvord AC, Sumi SM, Streissguth AP, Smith DW (1978) Brain malformations related to prenatal exposure to ethanol. J Pediatr 92:64–67

Collins RJ, Hermon BV, Gobé GC, Kerr JFR (1992) Internucleosomal DNA cleavage should not be the sole criterion for identifying apoptosis. Int J Radiat Biol 61:451–453

Dessens AB, Cohen-Kettenis PT, Mellenbergh GJ, Koppe JG, van de Poll NE, Boer K (2000) Association of prenatal phenobarbital and phenytoin exposure with small head size and with learning problems. Acta Paediatr 89:533–541

Dobbing J, Sands J (1979) The brain growth spurt in various mammalian species. Early Human Dev 3:79–84

Eckardt MJ, File SE, Gessa GL, Grant KA, Guerri C, Hoffman PL, Kalant H, Koob GF, Li TK, Taba-koff B (1998) Effects of moderate alcohol consumption on the central nervous system. Alcohol Clin Exp Res 22:998–1040

Famy C, Streissguth AP, Unis AS (1998) Mental illness in adults with fetal alcohol syndrome or fetal alcohol effects. Am J Psych 155:552–554

Grasl-Kraupp B Ruttkay-Nedecky B, Koudelka H, Bukowska K, Bursch W, Schulte-Hermann R (1995) In situ detection of fragmented DNA (TUNEL assay) fails to discriminate among apop-tosis, necrosis, and autolytic cell death: a cautionary note. Hepatology 21:1465–1468

Gwag BJ, Koh JY, Demaro JA, Ying HS, Jacquin M, Choi DW (1997) Slowly triggered excitotoxicity occurs be necrosis in cortical cultures. Neuroscience 77:393–401

Hoffman PL, Rabe CS, Moses F, Tabakoff B (1989) N-Methyl-D-aspartate receptors and ethanol: In-inhibition of calcium flux and cyclic GMP production. J Neurochem 52:1937–1940

Ikonomidou C, Qin Y, Labruyere J, Kirby C, Olney JW (1996) Prevention of trauma-induced neu-rodegeneration in infant rat brain. Pediatr Res 39:1020–1027

Ikonomidou C, Bosch F, Miksa M, Vöckler J, Bittigau P, Pohl D, Dikranian K, Tenkova T, Turski L, Olney JW (1999) Blockade of glutamate receptors triggers apoptotic neurodegeneration in the developing brain. Science 283:70–74

Ikonomidou C, Bittigau P, Ishimaru MJ, Wozniak DF, Koch C, Genz K, Price MT, Stefovska V, Hörster F, Tenkova T, Dikranian K, Olney JW (2000) Ethanol-induced apoptotic neurodegen-eration and fetal alcohol syndrome. Science 287:1056–1060

Ishimaru MJ, Ikonomidou C, Tenkova TI, Der TC, Dikranian K, Sesma MA, Olney JW (1999) Dis-tinguishing excitotoxic from apoptotic neurodegeneration in the developing rat brain. J Comp Neurol 408:461–476

Jevtovic-Todorovic V, Todorovic SM, Mennerick S, Powell S, Dikranian K, Benshoff N, Zorumski CF, Olney JW (1998) Nitrous oxide (laughing gas) is an NMDA antagonist, neuroprotectant and neurotoxin. Nature Med 4:460–463

Jones KL, Smith DW (1973) Recognition of the fetal alcohol syndrome in early infancy. Lancet ii:999–1001

Jones KL, Smith DW, Ulleland CN, Streissguth AP (1973) Pattern of malformation in offspring of chronic alcoholic mothers. Lancet I:1267–1271

Kerns KA, Don A, Mateer CA, Streissguth AP (1977) Implementation for a compensatory memory system in a school aged child with severe memory impairment. J Learn Disab 30:685–693

Kerr JFR, Wyllie AH, Currie AR (1972) Apoptosis: a basic biological phenomenon with wide-ranging implications in tissue kinetics. Br J Cancer 26:239–257

Lovinger DM, White G (1991) Ethanol potentiation of 5-hydroxytryptamine receptor-mediated ion current in neuroblastoma cells and isolated adult mammalian cells. Mol Pharmacol 40:263–270

Olney JW (1969) Brain lesions, obesity and other disturbances in mice treated with monosodium glutamate, Science 164:719–721

Olney JW, Wozniak DF, Price MT, Ikonomidou C (1998) Alcohol induces massive apoptotic neuro-degeneration in the developing rat CNS. Soc Neurosci Abst 24:1983

Pohl D, Bittigau P, Ishimaru MJ, Stadthaus D, Fuhr S, Voeckler J, Huebner C, Olney JW, Ikonomi-dou C (1999) Apoptotic cell death triggered by head injury in infant rats is potentiated by NMDA antagonists. Proc Natl Acad Sci 96:2508–2513

Sulik KK, Johnston MC, Webb MA (1981) Fetal alcohol syndrome: embryogenesis in a mouse model. Science 214:936–938

White G, Lovinger DM, Weight FF (1990) Ethanol inhibits NMDA-activated current but does not alter GABA-activated current in an isolated adult mammalian neuron. Brain Res 507:332–336

Zahn CA, Morrell MJ, Collins SD, Labiner DM, Yerby MS (1998) Management issues for women with epilepsy. Neurology 51:949–956

Apoptosis, Glial Cells and Parkinson's Disease

E. C. Hirsch

Summary

The glial reaction is generally considered to be a consequence of neuronal death in neurodegenerative diseases such as Alzheimer's disease, Huntington's disease and Parkinson's disease.

In Parkinson's disease, postmortem examination reveals that the loss of dopaminergic neurons in the substantia nigra is associated with massive astrogliosis and the presence of activated microglial cells. Recent evidence suggests that the disease may progress even when the initial cause of neuronal degeneration has disappeared, implying that toxic substances released by glial cells may be involved in neuronal degeneration. Glial cells can release various compounds, including pro-inflammatory cytokines. These substances may act on specific receptors, located on the dopaminergic neurons, that contain intracytoplasmic death domains and are involved in apoptosis. Alternatively, since cytokines are known to induce the expression of nitric oxide via the induction and activation of the low affinity IgE receptor CD23, the gradual release of nitric oxide from glial cells may account for the increased oxidative stress, protein nitration, altered iron homeostasis and blood vessel alterations reported in the disease. In turn, such cellular alterations may provoke the degeneration of dopaminergic neurons. The exact cascade of events leading to neuronal degeneration in Parkinson's disease is not known but may involve activation of proteases such as caspase-3, which are known effectors of the cascade of events leading to nerve cell death.

Introduction

Neurodegenerative diseases such as Alzheimer's disease, Huntington's disease or Parkinson's disease are characterized by a slow and progressive neuronal loss. Parkinson's disease is of particular interest as the pathological process specifically occurs in dopaminergic neurons. During the past decade, however, it has become clear that Parkinson's disease is not a disorder of all dopaminergic neurons. The loss of dopaminergic neurons is severe in the substantia nigra pars compacta (76% loss), intermediate in the substantia nigra pars lateralis (34% loss), the ventral tegmental area (55% loss) and the

Henderson/Green/Mariani/Christen (Eds.)
Neuronal Death by Accident or by Design
© Springer-Verlag Berlin Heidelberg 2001

peri- and retrorubral region (31% loss), and almost nil in the central gray substance (3% loss; Hirsch et al. 1988). Furthermore, a recent study indicated that the distribution of lesions in the substantia nigra pars compacta is highly heterogeneous in this disease. Thus, within the substantia nigra pars compacta, Damier et al. (1999a) identified dopamine-containing neurons in calbindin-rich regions (matrix) and in five calbindin-poor pockets (nigrosomes), defined by analyzing the three-dimensional networks formed by the calbindin-poor zones. The same authors showed that the degree of neuronal loss in Parkinson's disease was significantly higher in the nigrosomes than in the matrix (Damier et al. 1999b). Furthermore, a spatio-temporal analysis of the neuronal loss showed that the pathological process begins in the main pocket (nigrosome 1) and then spreads to other nigrosomes and the matrix along rostral, medial and dorsal axes of progression. Yet, although the distribution of lesions is now well established in this disease, the cause of neuronal loss is still unknown. In this context, postmortem studies of the brains of Parkinson's disease patients showed increased lipid peroxidation and iron content, decreased glutathione content, altered expression of antioxidant enzymes and impaired mitochondrial functions, suggesting a major role of oxidative stress in the neurodegenerative process (Agid 1991; Agid et al. 1993; Hirsch 1993; Hirsch and Faucheux 1998; Hunot et al. 1996; Jenner 1996, 1998; Jenner and Olanow 1996, 1998; Olanow 1992, 1997; Owen et al. 1996; Przedborski and Jackson-Lewis 1998; Qureshi et al. 1995). Yet, most of these changes also occur in other neurodegenerative diseases, suggesting that they represent a common pathway to a certain kind of neuronal death and not an initial cause of the disease. In line with this hypothesis, Langston and co-workers recently shown that, even in the absence of the primary cause of neurodegeneration, an ongoing cell loss can occur in parkinsonian syndromes, suggesting that a time-limited insult to the nigrostriatal system can set in motion a self-perpetuating process of neurodegeneration (Langston et al. 1999). In their study, these authors provided a neuropathological study of 1-methyl-4-phenyl-1,2,3,6-tetrahydropyridine (MPTP)-induced parkinsonism in humans. They showed that three subjects who self administered the drug under the impression that it was synthetic heroine and subsequently developed severe and unremitting parkinsonism displayed at autopsy, three to 16 years after intoxication, a still active neuronal loss characterized by a severe gliosis and clustering of microglial cells around neurons. Taken together, these data suggest that the glial cells may participate in the cascade of events leading to nerve cell death in neurodegenerative disorders such as Parkinson's disease. In this review, we will successively analyze the role of some putatively toxic factors secreted by glial cells in the degenerative process in parkinsonian syndromes.

Can Glial Cells Play a Deleterious Role for Dopaminergic Neurons in Parkinsonian Syndromes?

A deleterious role for glial cells in the parkinsonian syndromes was initially proposed in a pioneering study by McGeer and co-workers, in which they reported the presence of activated microglial cells in the substantia nigra of patients with Parkinson's disease displaying HLA-DR-immunoreactivity (McGeer et al. 1988). More recently, Damier and co-workers performed a quantitative analysis of glial cells and reported an increased density of astroglial cells, characterized by an immunoreactivity for the glial fibrillary acidic protein (GFAP) in the substantia nigra of patients with Parkinson's disease (Damier et al. 1993). Thus, it is evident that a microglial and astroglial reaction occurs in the substantia nigra of patients with Parkinson's disease. Although, the glial cells may have a neuroprotective effect for dopaminergic neurons, they could also secrete deleterious compounds, such as pro-inflammatory cytokines. Such a concept is supported by several in vitro and in vivo experiments. Indeed, lipopolysaccharides (LPS), which induce the expression of pro-inflammatory cytokines, have been shown to kill dopaminergic neurons in mixed neuron glial cell cultures but not in pure neuronal cultures (Bronstein et al. 1995). This indicates that the production of a glial factor induced by LPS (very likely cytokines) is able to kill dopaminergic neurons at least in vitro. These results were recently confirmed by McNaught and co-workers (1999) in experiments in which they showed that LPS-activated astrocytes could cause neuronal death in a time-dependent manner in primary ventral mesencephalic neuronal cultures. Furthermore, these authors extended this finding by showing that the toxicity of agents capable of inducing the death of dopaminergic neurons, such as MPP^+ or 6-hydroxydopamine, was enhanced when the dopaminergic neurons were cultured with LPS-activated astrocytes. Taken together, these data clearly show that activated glial cells can participate in the death of dopaminergic neurons. The factors involved in this deleterious effect are very likely cytokines, including tumor necrosis factor-a (TNF-a; Fig. 1). TNF-a exerts a dose-dependent toxicity specifically for dopaminergic neurons in primary mesencephalic cultures of rat mesencephalon (McGuire, et al., personal communication).

Furthermore, TNF-a may also play a deleterious role in vivo for dopaminergic neurons, as overexpression of this cytokine in the central nervous system in transgenic animals reduces tyrosine hydroxylase immunoreactivity in the striatum and the dorso-medial hypothalamic area, resulting in an impairment of grooming behavior (Aloe and Fiore 1997). Thus, experimental data obtained in vivo and in vitro suggest that pro-inflammatory cytokines may participate in the cascade of events leading to nerve cell death in pathologies of dopaminergic neurons. This concept is further supported by postmortem analysis showing an increased density of glial cells expressing TNF-a, interferon-γ and interleukin-1β in the substantia nigra of patients with Parkinson's disease (Boka et al. 1994; Hunot et al. 1999). In addition, the concentration of interleukin-2, 4 and 6 is also raised in ventricular cerebro-

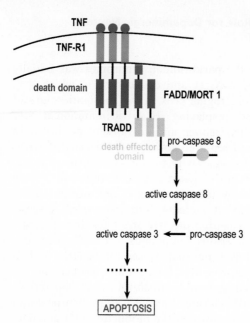

Fig. 1. TNFα receptor I proapoptotic transduction pathway

spinal fluid and in the caudate nucleus and putamen of patients with Parkinson's disease (Mogi et al. 1994a,b, 1996a,b). Taken as a whole, the data reviewed here suggest that dopaminergic neurons in the substantia nigra of patients with Parkinson's disease are surrounded by a high density of glial cells secreting pro-inflammatory cytokines and that, at least in in vivo and in vitro experimental models, these cytokines may play a deleterious role for dopaminergic neurons. In the next section, we will discuss the mechanisms by which pro-inflammatory cytokines could exert a deleterious effect.

By what Mechanisms Could Pro-Inflammatory Cytokines Play a Deleterious Role in Parkinson's Disease?

Pro-inflammatory cytokines may damage dopaminergic neurons by secreting factors that either diffuse into the neurons or act on specific receptors located on the neuronal membranes. These two mechanisms, which are not mutually exclusive, will be discussed successively.

Recent evidence implies an involvement of cytokines in the activation of nitric oxide production in glial cells (Hunot et al. 1999). Given that nitric oxide is a diffusible compound that may penetrate into the neurons and exert toxic effects, this mechanism may account, at least in part, for the deleterious role of glial cells. The mechanism by which cytokines could induce nitric oxide production is now quite well understood. Cytokines, and more specifically interferon-γ, induce the expression of a molecule called CD23. CD23 is a low-affinity immunoglobulin E receptor consisting of a 45 kDa, transmembrane type-II glycoprotein and is a member of the C-type lectine family (Delespesse et al. 1991). It is expressed at the cell surface of various cell types after stimulation by cytokines (Dugas et al. 1995). Recently, cytokine induction of CD23 was reported in an astrocytoma cell line (Hunot et al. 1999). Although interferon-γ alone was sufficient to induce CD23 expression, the addition of interleukin-1β and TNFα increased the level of expression. The presence of CD23 alone, however, was not sufficient to induce the production of nitric oxide by astrocytes. Appropriate ligands were also required to activate the inducible form of nitric oxide synthase (also known as nitric oxide synthase II). Indeed, CD23 ligation induced the expression of both nitric oxide synthase II mRNA, detected by reverse transcriptase polymerase chain reaction (RT-PCR), and protein, detected by Western blot. In addition, the nitric oxide produced by nitric oxide synthase could stimulate the production of pro-inflammatory cytokines, such as TNFα and interleukin-6 (Arock et al. 1994; Hunot et al. 1999; Mossalayi et al. 1994). Together, these data suggest that cytokines could be involved in the induction of nitric oxide and the subsequent release of nitric oxide in the central nervous system. The elevated levels of cytokines observed in the brains of patients with Parkinson's disease may therefore increase the production of nitric oxide during the pathological process. In agreement with this, an increase in the density of CD23-positive glial cells has been reported in the substantia nigra of patients with Parkinson's disease (Hunot et al. 1999). Similarly, the increased number of glial cells with nitric oxide synthase II and nicotinamide adenine dinucleotide phosphate (NADPH) diaphorase activity in the substantia nigra of patients with Parkinson's disease suggests that CD23 does indeed stimulate the expression of nitric oxide synthase (Hunot et al. 1996).

Whether nitric oxide is involved in nerve cell death in Parkinson's disease remains to be determined. However, experimental evidence suggests that this may be the case. Indeed, the concentration of nitrates is increased in the cerebrospinal fluid of patients with Parkinson's disease (Qureshi et al. 1995), and 3-nitro-tyrosine, an index of protein nitrosation induced by nitric oxide-derived oxidizing molecules peroxinitrates, has been detected in nigral dopaminergic neurons (Good et al. 1998) of patients with Parkinson's disease. The mechanisms by which nitric oxide may be deleterious to dopaminergic neurons is not known. It is, however, known that nitric oxide combined with superoxide radicals can form peroxinitrates that are extremely toxic to cells. Alternatively, it has been shown that nitric oxide can release iron from the iron-buffering protein, ferritin (Juckett et al. 1998). As free iron is extremely toxic and potentiates the formation of hydroxyl radicals, it may thus contribute to the toxic effect of nitric oxide. In line with this, a deregulation of ferritin expression possibly induced by nitric oxide has recently been observed in the substantia nigra of patients with Parkinson's disease (Faucheux et al., submitted for publication). These data suggest that the cytokines produced by glial cells may reinforce oxidative stress by producing nitric oxide and increasing the level of free iron. In line with this finding, experimental data in animal models of Parkinson's disease in which the production of nitric oxide has been reduced by invalidation of the inducible form of nitric oxide synthase show that nitric oxide may participate in the death of dopaminergic neurons (Liberatore et al. 1999; Dehmer et al. 2000). Similarly, pretreatment of animals by intraventricular injection with a relatively selective prototype iron chelator, desferioxamine, attenuates the 6-hydroxydopamine lesion of nigrostriatal neurons in rats, suggesting that iron may also participate in the death of dopaminergic neurons (Youdim et al. 1993).

Yet, these data do not rule out the possibility that TNFα or other cytokines may participate more directly in the death of dopaminergic neurons by acting on specific receptors located at their plasmic membrane level. TNFα may also play a more direct deleterious role for dopaminergic neurons by activating pro-apoptotic pathways coupled to TNFα receptors. Two types of TNFα receptors, TNFR1 (TNFR p55) and TNFR2 (TNFR p75), have been described in the central nervous system. These receptors display only moderate sequence homology (20%) in their extra-cellular domain, and no homology in their intra-cellular portion. Furthermore, only TNFR1 displays in its intra-cytoplasmic loop a death domain that is characteristic for the TNFα receptors super-family (Baker and Reddy 1998). The expression of these two receptors is also differentially regulated. Thus, TNFR1 is expressed constitutively at a very low level, whereas TNFR2 is induced by extrinsic factors. Furthermore, TNFR2 binds TNFα at lower concentrations than TNFR1. This variability in their expression level and TNFα binding capacity may explain, at least in part, the different physiological responses of these two receptors (Vandenabeele et al. 1994). As far as dopaminergic neurons are concerned, TNFR1 is expressed by the dopaminergic neurons of the human substantia nigra, suggesting that TNFα secreted by the surrounding glial cells may activate pro-apoptic pathways in Parkinson's disease (Boka et al. 1994). By analogy to the immune system, TNFα could induce a trimerisation of TNFR1 upon binding. This process may allow the adaptor proteins TRADD and FADD to be recruited via the intracytoplasmic death domain of TNFR1. This hypothesis is supported by the fact that dopaminergic neurons of the human substantia nigra express these adaptor molecules, suggesting that they may be recruited following stimulation of the TNFR1 receptor (Hartmann et al. 2000). The recruitment of these adaptor molecules may in turn lead to the auto-proteolytic activation of a family of proteases called caspases.

In line with this, a recent postmortem study showed a dramatic increase in the number of dopaminergic neurons expressing activated caspase-8 in the substantia nigra of patients with Parkinson's disease compared to control subjects (Hartmann et al. 2001). Furthermore, caspase-3, a major effector of this proapoptotic pathway, has also been shown to be activated in the substantia nigra of patients with Parkinson's disease (Hartmann et al. 2000). This finding suggests that molecular pathways known to be involved in several models of neuronal apoptosis in vitro and in vivo may participate in the apoptotic death of dopaminergic neurons in Parkinson's disease.

Several indepedent groups of investigators have reported the presence of apoptotic cells in the parkinsonian substantia nigra, although this presence has not been observed by other investigators (for review, see Hartmann and Hirsch 2001). Mochizuki et al. (1996) were the first to describe a significantly higher number of TUNEL-positive neurons in the brains of patients with Parkinson's disease compared to controls. Using other techniques, such as electron microscopy (Anglade et al. 1997), or a combination of end-labeling techniques or fluorescent dyes, other investigators (Tatton et al. 1998; Tompkins et al. 1997) confirmed the presence of apoptotic cells in the parkinsonian substantia nigra. Yet, other studies did not find evidence of apoptotic dopaminergic neurons in Parkinson's disease, but rather suggested that peri-mortem factors may interfere with the results (Kingsbury et al. 1998; Wullner et al. 1999).

Thus, since the significance of purely morphological human postmortem features suggestive of apoptosis has remained controversial, the investigations of molecular apoptotic markers such as activated forms of caspases in Parkinson's disease brains provide strong evidence in favor of an apoptotic mode of cell death in Parkinson's disease. Nevertheless, this does not imply that all dopaminergic neurons degenerate by apoptosis, since electronic microscopy studies also showed the presence of dopaminergic neurons displaying the morphological features of autophagic degeneration (Anglade et al. 1997). The results of postmortem studies also need to be treated with the utmost caution, as it is difficult to determine whether changes observed postmortem in the brain, such as activation of caspases, are directly associated with the induction of neuronal degeneration or are merely a consequence of it. This issue was addressed in several experimental models which showed, either in vitro in primary dopaminergic cultures of the rat embryo mesencephalon or in mice intoxicated by MPTP, that caspase-8 and caspase-3 activation strictly follows the time course of apoptosis development (Hartmann et al. 2000, 2001; Turmel et al. 2001).

Thus, if caspases are indeed involved in the death of dopaminergic neurons in parkinsonian syndromes, the key question is raised as to whether inhibition of caspases can prevent the loss of dopaminergic neurons in parkinsonian syndromes. This hypothesis has been tested in primary mesencephalic cultures of the rat mesencephalon, in which MPP$^+$, the active metabolite of MPTP, has been shown to induce apoptosis. Nevertheless, cultures treated with zVAD-fmk, a broad spectrum caspase inhibitor with relative specificity for caspases 2, 3 and 7, and/or zIETD-fmk, with relative specificity for caspase-8, did not survive MPTP$^+$ intoxication (Hartmann et al. 2001). Furthermore, co-treatment with caspase inhibitors and MPP$^+$ induced an even more dramatic decrease in the number of dopaminergic neurons than treatment with MPP$^+$ alone. These differences could not be attributed to the caspase inhibitors, as cultures treated alone with these compounds were not significantly altered. In addition, a careful analysis of the mode of cell death induced by MPP$^+$ in the presence of the caspase inhibitors showed that the latter compounds induced a switch from apoptosis to another type of cell death, characterized by an absence of chromatin condensation, alteration of processes and finally a complete destruction of the cell bodies, resembling the morphological criteria for necrosis. As necrosis predominantly occurs in states of intracellular energy depletion (Nicotera et al. 2000), this switch from apoptosis to necrosis may have been provoked by a reduced energy metabolism. Indeed, whereas cultures treated with MPP^+ exhibited a slight but not significant decrease in ATP levels (an index of energy resources), ATP levels decreased significantly in cultures treated with zVAD-fmk or zIETD-fmk. In addition, when MPP^+ intoxicated cultures were co-treated with caspase inhibitors and glucose, an energy source, the deleterious effect of MPP^+ on cell survival was reduced compared to cultures treated without the addition of glucose. However, under such conditions, caspase inhibition did not affect [^{3}H] dopamine uptake, taken as an index of neuronal functionality.

It has been suggested that, while caspase inhibition may block the morphological manifestation of apoptosis, cell viability and functionality are not affected in cells morphologically preserved by inhibition of the apoptotic cascade (Werth et al. 2000). These data show the need for a cautious approach to the use of caspase inhibitors in the treatment of Parkinson's disease. First, intracellular ATP depletion following caspase inhibition and respiratory failure in PD may cause a switch from apoptosis to necrosis. Second, necrosis induced by caspase inhibitors may also have deleterious effects on the surrounding tissue and contribute to further damage of adjacent neurons. Third, even if intracellular energy levels are raised, the ensuing morphological preservation does not restore neuronal functionality.

Conclusion

The data reviewed her support the notion that glial cells and, more generally, the pro-inflammatory cytokines secreted by these cells, may be involved in the progression of neuronal loss in Parkinson's disease. Nevertheless, while it cannot be excluded that such a phenomenon may participate in the primary cause of cell death, this hypothesis is unlikely to be validated, and the production of cytokines is rather one event in the cascade of events leading to neuronal degeneration. Indeed, a glial reaction and secretion of pro-inflammatory cytokines have been reported in other neurodegenerative disorders, such as Alzheimer's disease or even in encephalopathy associated with AIDS. Nevertheless, this phenomenon may be involved in the perpetuation of neuronal loss in neurodegenerative disorders. From a therapeutic point of view, therapeutic intervention aimed at blocking this phenomenon may prove to be beneficial for the survival of dopaminergic neurons, as could be the case in Alzheimer's disease. Nevertheless, the data presented here suggest that a single drug such as a caspase inhibitors is unlikely to be effective. Future therapeutic strategies might be based on a combination of drugs, such as caspase inhibitors, energy providers and neurotrophic factors. Thus, a multiple therapy may be necessary to slow down or even block the death of dopaminergic neurons in parkinsonian syndromes. Such a concept has already been developed with quite good results in other pathologies such as AIDS.

References

Agid Y (1991) Parkinson's disease: pathophysiology. Lancet (Ed. Française) 337:1311–1324

Agid Y, Ruberg M, Javoy-Agid F, Hirsch E, Raisman-Vozari R, Vyas S, Faucheux B, Michel P, Kastner A, Blanchard V, Damier P, Villares J, Ping Zhang (1993) Are dopaminergic neurons selectively vulnerable to Parkinson's disease? Adv Neurol 60:148–164

Aloe L, Fiore M (1997) TNF-α expressed in the brain of transgenic mice lowers central tyrosine hydroxylase immunoreactiviy and alters grooming behavior. Neurosci Lett 238:65–68

Anglade P, Vyas S, Javoy-Agid F, Herrero MT, Michel PP, Marquez J, Mouatt-Prigent A, Ruberg M, Hirsch EC, Agid Y (1997) Apoptosis and autophagy in nigral neurons of patients with Parkinson's disease. Histol Histopathol 12:25–31

Arock M, Le Goff L, Bécherel PA, Dugas B, Debré P, Mossalayi MD (1994) Involvement of FcepsilonRII/CD23 and L-arginine dependent pathway in IgE-mediated activation of human eosinophils. Biochem Biophys Res Commun 203:265–271

Baker SJ, Reddy EP (1998) Modulation of life and death by the TNF receptor superfamily. Oncogene 17:3261–3270

Boka G, Anglade P, Wallach D, Javoy-Agid F, Agid Y, Hirsch EC (1994) Immunocytochemical analysis of tumor necrosis factor and its receptors in Parkinson's disease. Neurosci Lett 172:151–154

Bronstein DM, Perez-Otano I, Sun V, Mullis Sawin SB, Chan J, Wu GC, Hudson PM, Kong LY, Hong JS, McMillian MK (1995) Glia-dependent neurotoxicity and neuroprotection in mesencephalic cultures. Brain Res 704:112–116

Damier P, Hirsch EC, Zhang P, Agid Y, Javoy-Agid F (1993) Glutathione peroxidase, glial cells and Parkinson's disease. Neuroscience 52:1–6

Damier P, Hirsch EC, Agid Y, Graybiel AM (1999a) The substantia nigra of the human brain: I. Nigrosomes and the nigral matrix, a compartmental organization based on calbindin D-28K immunohistochemistry. Brain 122:1421–1436

Damier P, Hirsch EC, Agid Y, Graybiel AM (1999b) The substantia nigra of the human brain: II. Patterns of loss of dopamine-containing neurons in Parkinson's disease. Brain 122:1437–1448

Dehmer T, Lindenau J, Haid S, Dichgans J, Schulz JB (2000) Deficiency of inducible nitric oxide synthase protects against MPTP toxicity in vivo. J Neurochem 74:2213–2216

Delespesse G, Sutter U, Mossalayi MD, Bettler B, Sarfati M, Hofstetter H, Kilchherr E, Debré P, Dalloul AH (1991) Expression, structure, and function of the CD23 antigen. Adv Immunol 49:149–191

Dugas B, Mossalayi MD, Damais C, Kolb J-P (1995) Nitric oxide production by human monocytes: evidence for a role of CD23. Immunol Today 16:574–580

Good PF, Hsu A, Werner P, Perl DP, Olanow CW (1998) Protein nitration in Parkinson's disease. J Neuropathol Exp Neurol 57:338–342

Hartmann A, Hunot S, Michel PP, Muriel M-P, Vyas S, Faucheux BA, Mouatt-Prigent A, Turmel E, Evan GI, Agid Y, Hirsch EC (2000) Caspase-3: a vulnerability factor and final effector in apoptotic death of dopaminergic neurons in Parkinson's disease. Proc Natl Acad Sci 97:2875–2880

Hartmann A, Hirsch EC (2001) Parkinson's disease: the apoptosis hypothesis revisited. Adv Neurol 86:143–153

Hirsch EC (1993) Does oxidative stress participate in nerve cell death in Parkinson's disease? Eur Neurol 33:52–59

Hirsch EC, Faucheux BA (1998) Iron metabolism and Parkinson's disease. Mov Disord 13(S1):39–45

Hirsch EC, Graybiel AM, Agid Y (1988) Melanized dopaminergic neurons are differentially susceptible to degeneration in Parkinson's disease. Nature 334:345–348

Hunot S, Boissière F, Faucheux B, Brugg B, Mouatt-Prigent A, Agid Y, Hirsch EC (1996) Nitric oxide synthase and neuronal vulnerability in Parkinson's disease. Neuroscience 72:355–363

Hunot S, Dugas N, Faucheux B, Hartmann A, Tardieu M, Debré P, Agid Y, Dugas B, Hirsch EC (1999) Fc(epsilon)RII/CD23 is expressed in Parkinson's disease and induces, in vitro, production of nitric oxide and tumor necrosis factor-alpha in glial cells. J Neurosci 19:3440–3447

Jenner P (1996) Oxidative stress in Parkinson's disease and other neurodegenerative disorders. Pathol Biol 44:57–64

Jenner P (1998) Oxidative mechanisms in nigral cell death in Parkinson's disease. Mov Disord 13:S24–S34

Jenner P, Olanow CW (1996) Oxidative stress and the pathogenesis of Parkinson's disease. Neurology 47:S161–S170

Jenner P, Olanow CW (1998) Understanding cell death in Parkinson's disease. Ann Neurol 44:S72–S84

Juckett M, Zheng Y, Yuan H, Pastor T, Antholine W, Weber M, Vercellotti G (1998) Heme and the endothelium. Effects of nitric oxide on catalytic iron and heme degradation by heme oxygenase. J Biol Chem 273:288–297

Kingsbury AE, Marsden CD, Foster OJ (1998) DNA fragmentation in human substantia nigra: apoptosis or perimortem effect? Mov Disord 13:877–884

Langston JW, Forno LS, Tetrud J, Reeves AG, Kaplan JA, Karluk D (1999) Evidence of active nerve cell degeneration in the substantia nigra of humans years after 1-methyl-4-phenyl-1,2,3,6-tetrahydropyridine exposure. Ann Neurol 46:598–605

Liberatore GT, Jackson-Lewis V, Vukosavic S, Mandir AS, Vila M, McAuliffe WG, Dawson VL, Dawson TM, Przedborski S (1999) Inducible nitric oxide synthase stimulates dopaminergic neurodegeneration in the MPTP model of Parkinson's disease. Nature Med 5:1403–1409

McGeer PL, Itagaki S, Boyes BE, McGeer EG (1988) Reactive microglia are positive for HLA-DR in the substantia nigra of Parkinson's and Alzheimer's disease brains. Neurology 38:1285–1291

McNaught KS, Jenner P (1999) Altered glial function causes neuronal death and increases neuronal susceptibility to 1-methyl-4-phenylpyridinium and 6-hydroxydopamine-induced toxicity in astrocytic/ventral mesencephalic co-cultures. J Neurochem 73:2469–2476

Mochizuki H, Goto K, Mori H, Mizuno Y (1996) Histochemical detection of apoptosis in Parkinson's disease. J Neurol Sci 137:120–123

Mogi M, Harada M, Riederer P, Narabayashi H, Fujita K, Nagatsu T (1994a) Tumor necrosis factor-α (TNF-α) increases both in the brain and in the cerebrospinal fluid from parkinsonian patients. Neurosci Lett 165:208–210

Mogi M, Harada M, Kondo J, Riederer P, Inagaki H, Minami M, Nagatsu T (1994b) Interleukin-1 β, interleukin-6, epidermal growth factor and transforming growth factor-α are elevated in the brain from parkinsonian patients. Neurosci Lett 180:147–150

Mogi M, Harada M, Kondo T, Riederer P, Nagatsu T (1996a) Interleukin-2 but not basic fibroblast growth factor is elevated in parkinsonian brain. J Neural Transm 103:1077–1081

Mogi M, Harada M, Narabayashi H, Inagaki H, Minami M, Nagatsu T (1996b) Interleukin (IL)-1 β, IL-2, IL-4, IL-6 and transforming growth factor-α levels are elevated in ventricular cerebrospinal fluid in juvenile parkinsonism and Parkinson's disease. Neurosci Lett 211:13–16

Mossalayi MD, Paul-Eugène N, Ouaaz F, Arock M, Kolb J-P, Kilchherr E, Debré P, Dugas B (1994) Involvement of FcepsilonRII/CD23 and L-arginine-dependent pathway in IgE-mediated stimulation of human monocyte functions. Int Immunol 6:931–934

Nicotera P, Leist M, Fava E, Berliocchi L, Volbracht C (2000) Energy requirement for caspase activation and neuronal cell death. Brain Pathol 10:276–282

Olanow CW (1992) An introduction of the free radical hypothesis in Parkinson's disease. Ann Neurol 32:S2–S9

Olanow CW (1997) Attempts to obtain neuroprotection in Parkinson's disease. Neurology 49:S26–S33

Owen AD, Schapira AH, Jenner P, Marsden CD (1996) Oxidative stress and Parkinson's disease. Ann NY Acad Sci 786:217–223

Przedborski S, Jackson-Lewis V (1998) Experimental developments in movement disorders: update on proposed free radical mechanisms. Curr Opin Neurol 11:335–339

Qureshi GA, Baig S, Bednar I, Sodersten P, Forsberg G, Siden A (1995) Increased cerebrospinal fluid concentration of nitrite in Parkinson's disease. NeuroReport 6:1642–1644

Tatton NA, Maclean-Fraser A, Tatton WGPDP, Olanow CW (1998) A fluorescent double-labeling method to detect and confirm apoptotic nuclei in Parkinson's disease. Ann Neurol 44:S142–S148

Tompkins MM, Basgall EJ, Zamrini E, Hill WD (1997) Apoptotic-like changes in Lewy-body-associated disorders and normal aging in substantia nigral neurons. Am J Pathol 150:119–131

Turmel H, Hartmann A, Parain K, Douhou A, Srinivasan A, Agid Y, Hirsch EC (2001) Caspase-3 activation in 1-methyl-4-phenyl-1,2,3,6-tetrahydropyridine (MPTP)-treated mice. Mov Disord 16:185–189

Vandenabeele P, Declercq W, Beyaert R, Fiers W (1994) Two tumor necrosis factor receptors: structure and function. Trends Cell Biol 5:392–399

Werth JL, Deshmukh M, Cocabo J, Johnson EM Jr, Rothman SM (2000) Reversible physiological alterations in sympathetic neurons deprived of NGF but protected from apoptosis by caspase inhibition or Bax deletion. Exp Neurol 161:203–211

Wullner U, Kornhuber J, Weller M, Schulz JB, Loschmann PA, Riederer P, Klockgether T (1999) Cell death and apoptosis regulating proteins in Parkinson's disease: a cautionary note. Acta Neuropathol (Berl) 97:408–412

Youdim MB, Ben Shachar D, Riederer P (1993) The possible role of iron in the etiopathology of Parkinson's disease. Mov Disord 8:1–12

Mechanisms of Neuronal Death:
An in vivo Study in the Lurcher Mutant Mice

F. Selimi, A. Campana, J. Bakouche, A. Lohof, M. W. Vogel, and J. Mariani

Summary

Lurcher is a gain-of-function point mutation located in the gene encoding the δ_2 subunit of glutamate receptors (GRID2). The Lurcher mutation is lethal when homozygous. Heterozygous mice are ataxic due to a massive neuronal loss in their cerebellum. Lurcher Purkinje cells expressing the mutated allele are depolarized and die from the second postnatal week onwards, suggesting an excitotoxic process. Target-related cell death affects more than 90% of granule cells and 60–75% of olivary neurons, the two Purkinje cell afferences.

Thus, the Lurcher heterozygous mouse is an ideal model to study in vivo the mechanism of two types of neuronal death: an excitotoxicity-like process and target-related neuronal death. The timing of Purkinje cell death onset in Lurcher mice, around P10, is concomitant with the beginning of synaptogenesis between parallel fiber and Purkinje cell, suggesting a potential role of granule cell-Purkinje cell interaction in the timing of Purkinje cell death. X-irradiation of Lurcher mice during granule cell genesis is a means to reduce granule cell number. In these mice, Purkinje cells degenerate with the same timing as in Lurcher controls, suggesting that granule cells do not influence this process although they differentiate surprisingly better than in non-irradiated mutants.

The molecular cascade leading to apoptosis, a particular type of cell death, has been well defined, especially in vitro. Two families of proteins have an essential role in the regulation of apoptosis: the Bcl-2 family and the caspases. TUNEL-labeling studies have suggested the involvement of apoptosis in both types of neuronal death affecting the Lurcher nervous system.

The caspases are structurally similar cystein proteases that cleave their substrates specifically after an aspartate residue. They are synthesized as a precursor that is activated by cleavage, resulting in the formation of a large and a small subunit. Two heterodimers then associate to form the final active protease. Three categories of caspases can be distinguished by the specificity of their substrate cleavage site: caspases generating mature proinflammatory cytokines (caspase-1, -4, -5), effector caspases (caspase-2, -3, -7 and CED3) and initiator caspases (caspase-6, -8, -9) of apoptosis. The analysis of caspase-3 expression has shown an up-regulation of pro-caspase-3 in 25% of Purkinje cells of Lurcher mice. Activation of this caspase was also detected

Henderson/Green/Mariani/Christen (Eds.)
Neuronal Death by Accident or by Design
© Springer-Verlag Berlin Heidelberg 2001

in 1–3% of these cells, as was TUNEL labeling. Dying granule cells and olivary neurons also contained activated caspase-3. These results further suggest the involvement of an apoptotic process in the two types of neuronal death occurring in Lurcher mice.

The Bcl-2 family contains both pro- and anti-apoptotic members. Overexpression of the anti-apoptotic protein Bcl-2 in Lurcher mice does not rescue Purkinje cells but is able to delay this process, as Purkinje cells can still be found in two-month-old Lurcher mice overexpressing Bcl-2. We further analyzed the involvement of the Bcl-2 family by studying the role of the proapoptotic protein Bax. Bax up-regulation has been shown in both Purkinje cells and granule cells in Lurcher mice. To analyze the effect of Bax inactivation, we generated double-mutants, i.e. Bax knock-out Lurcher mice. One-month-old animals had a 40% increase in granule cell number. This increase was still observed in two-month-old animals, showing that Bax inactivation persistently inhibited the target-related death of granule cells. However, olivary neuron degeneration was not prevented in Bax knock-out Lurcher mice, showing that Bax involvement in target-related cell death depends on the neuronal population. In one-month-old animals, Purkinje cell number was the same in Bax knock-out Lurcher mice and in Lurcher controls. However, an increased number of Purkinje cells is detected in P15 Bax knock-out Lurcher mice. Thus, Bax inactivation is not sufficient to inhibit Purkinje cell death induced by the Lurcher mutation, but it is able to delay this process for a short period. In the Bax knock-out Lurcher mice, caspase-3 activation is inhibited in both Purkinje cells and granule cells, whereas pro-caspase-3 up-regulation in Purkinje cells is not influenced. Granule cell rescue in this model can be correlated to the inhibition of caspase-3 activation. Interestingly, the inhibition of caspase-3 activation is not sufficient to rescue Purkinje cells, suggesting that another pathway, for example another caspase or a caspase-independent mechanism, is able to mediate Lurcher Purkinje cell death.

The study of Lurcher mice highlights the point that different pathways underlie neuronal death depending on the death stimuli and also on the neuronal population. Different proteins in granule cells and olivary neurons mediate target-related neuronal death. Moreover, Lurcher Purkinje cells express different apoptosis-inducing molecules, activated caspase-3 and Bax, but are not rescued by their inhibition, suggesting that in one cell type several pathways of cell death can be induced by one stimulus.

Introduction

The heterozygous Lurcher mutant ($Grid2^{Lc/+}$) has been studied extensively as a model for understanding the mechanisms of cell-autonomous and target-related neuronal cell degeneration. In the $Grid2^{Lc/+}$ mutant, almost all cerebellar Purkinje cells, 60 to 75% of the olivary neurons, and 90% of the granule cells degenerate starting after the first week of postnatal development

(Phillips 1960; Caddy and Biscoe 1979). Studies of $Grid2^{Lc/+} \leftrightarrow$ wild type chimeras established that $Grid2^{Lc/+}$ Purkinje cell death is cell autonomous, whereas granule cell and olivary neuron cell death is secondary to the loss of their primary target, the Purkinje cells (Wetts and Herrup 1982 a,b). The semi-dominant Lurcher mutation (*Lc*) is lethal when homozygous and has been identified as a point mutation in the δ_2 glutamate receptor gene (*Grid2*) located on chromosome 6 (Zuo et al. 1997). The GRID2 protein is expressed specifically in cerebellar Purkinje cells (Araki et al. 1993) and has an unspecified function. The G-to-A transition found in the *Lc* allele is a gain-of-function mutation, resulting in a large, constitutive inward current in $Grid2^{Lc/+}$ Purkinje cells, suggesting that the cell dies by excitotoxicity (Zuo et al. 1997; Kohda et al. 2000). It has been proposed that the $Grid2^{Lc/+}$ Purkinje cell dies by an apoptotic mechanism (Norman et al. 1995; Wullner et al. 1995, 1998). However, previous morphological studies have concluded that the $Grid2^{Lc/+}$ Purkinje cell death is necrotic (Dumesnil-Bousez and Sotelo 1993).

Thus, the precise nature of the mechanism underlying the $Grid2^{Lc/+}$ Purkinje cell death remains unclear and needs to be resolved by the analysis of other characteristic features of apoptosis and necrosis. These types of cell death were originally distinguished by morphological features. Apoptosis is now also characterized by the involvement of two families of proteins: the BCl-2 family and the caspases.

Caspases are structurally similar, aspartate-specific cystein proteases. They are synthesized as a precursor, which is activated by cleavage. Three categories of caspases can be distinguished by the specificity of their substrate cleavage site (Nicholson and Thornberry 1997; Thornberry et al. 1997): caspases generating mature pro-inflammatory cytokines (caspase-1, -4, -5), effector caspases (caspase-2, -3, -7 and CED3) and initiator caspases (caspase-6, -8, -9) of apoptosis. More specifically, caspase-3 activation has been demonstrated in several models of neurodegeneration in vitro (Du et al. 1997; Marks et al. 1998). Caspase-3 is also necessary for developmental neuronal death, as caspase-3 knock-out mice have prominent protrusions of brain tissue due to the presence of supernumerary neurons in many regions of their brain (Kuida et al. 1996).

The *bcl-2* proto-oncogene encodes an integral membrane protein that inhibits apoptosis in many cell types, although its mechanism of action is still not completely understood (Adams and Cory 1998). One important function of BCL-2 may be to counteract the pro-apoptotic effects of BAX (Gross et al. 1999). BCL-2 will bind to BAX and cell death may be regulated by the ratio of BCL-2 to BAX (Oltvai et al. 1993). During apoptosis, BAX induces the release of cytochrome *c* from the mitochondria, which in turn initiates a cascade of activation of the effector caspases (Desagher et al. 1999; Gross et al. 1999). BAX is involved in developmental cell death since deletion of the *Bax* gene reduces the incidence of naturally occurring neuronal death in vivo (White et al. 1998). $Bax^{-/-}$ neurons also survive trophic factor withdrawal in vitro or after in vivo axotomy (Deckwerth et al. 1996).

p53 is a tumor suppressor gene involved in neuronal death induced by various stimuli (Hughes et al. 1997) and is able to induce BAX-mediated caspase-3 activation and apoptosis in neurons (Cregan et al. 1999).

To analyze the molecular pathways leading to neuronal death in vivo, we have studied the expression and localization of caspase-3 in the cerebellum of $Grid2^{Lc/+}$ mice and determined the effects of deleting the pro-apoptotic genes *Bax* and p53 on Purkinje, granule and olivary neurons survival in the $Grid2^{Lc/+}$ mutant.

Materials and Methods

Animals and Genotyping

Lurcher heterozygotes ($Grid2^{Lc/+}$) and their control wild-type ($Grid2^{+/+}$) littermates were obtained by breeding $Grid2^{Lc/+}$ males with $Grid2^{+/+}$ females. $Grid2^{Lc/+}$;$Bax^{-/-}$ double mutants were generated by heterozygous matings of $Grid2^{Lc/+}$ and $Bax^{+/-}$ mutants in the animal facilities at the Université P. & M. Curie. Heterozygous $Bax^{+/-}$ mutants were obtained from Dr. Stanley Korsmeyer (Knudson et al. 1995). In the mating scheme to generate homozygous and heterozygous double mutants. $Grid2^{Lc/+}$ males were first mated with *Bax* heterozygous knock-out females. The $Grid2^{Lc/+}$;$Bax^{+/-}$ animals were then intercrossed. Similarly, p53 knock-out Lurcher mice were generated using p53$^{-/-}$ mice (kindly provided by Dr. Moshe Yaniv). Control animals used in this study were obtained in the same litters as the mutants. All experiments were performed on at least three animals of each genotype.

$Grid2^{Lc/+}$ mice were recognized by their ataxia from P15 onwards. The $Grid2^{Lc}$ allele was detected by PCR followed by Single Strand Chain Polymorphism (SSCP) as previously described (Zuo et al. 1997).

Genotyping for *Bax* and p53 was performed by Polymerase Chain Reaction (PCR) using a set of appropriate primers. PCR products were resolved on a 1.5% agarose gel.

Immunohistochemistry

Animals at P12, P15 and P20 were anesthetized by an intraperitoneal (IP) injection of 3.5% chloralhydrate (0.1 ml/g weight) and perfused with 0.9% NaCl, followed by 95% ethanol. Brains were dissected, incubated overnight in Clarke's fixative, and processed for paraffin-embedding. Ten-µm thick tissue sections (either parasagittal or coronal) were processed for double immunostaining using the anti-hamster caspase-3 antiserum and an anti-calbindin monoclonal antibody (CL-300, Sigma). Non-specific binding sites were blocked using 4% normal horse serum and 4% normal goat serum diluted in TBS, 0.2% Triton X-100. The sections were then incubated overnight with the anti-hamster caspase-3 antibody (dilution 1/150 in TBS, 0.2% Triton X-100) and CL-300 antibody (1/200). Caspase-3 immunolabelling was detected using an FITC-conjugated anti-rabbit antibody (Jackson ImmunoResearch), and calbindin immunolabelling using a Cy3-conjugated anti-mouse antibody (Jackson ImmunoResearch). In each experiment, control sections were included in which the anti-caspase-3 antibody was replaced by normal rabbit serum (dilution 1/150 in TBS, 0.2% Triton x-100). The percentage of Purkinje cells overexpressing pro-caspase-3 was determined by counting the number of calbindin-immunopositive Purkinje cells per section and the number of FITC-labeled Purkinje cells on the same sections (three sections per cerebellum).

To detect active caspase-3, immunohistochemistry was performed on brain sections using a rabbit polyclonal antibody directed again a peptide contained in the C-terminal part of the p20 subunit of caspase-3 (CM1 antibody, a kind gift of Dr. Srinivasan). This antibody has been demonstrated to recognize the p20 subunit of active caspase-3 (Namura et al. 1998). Ten-µm parasagittal sections were incubated overnight at 4°C using CM1 antibody (dilution 1/1000) according to a protocol previously described (Namura et al. 1998). Immunocomplexes were then revealed using an anti-rabbit biotinylated antibody and DAB-coupled ABC kit (Vectastain Elite ABC kit, Vector Laboratories Inc., Burlingame, CA). In each experiment, control sections were included in which either the primary or the secondary antibody was omitted. Adjacent sections stained with cresyl-violet were used to quantify the total number of Purkinje cells per section. To test for the presence of immunolabelled cells in the inferior olive, fourteen µm coronal sections of the brainstem were immunostained using the same protocol. Given the small number of positive cells, counts were performed on at least 15 sections per animal and results were expressed as the number of positive cells per 10 sections.

DNA Fragmentation Detection and Double-Labelling with CM1 Antibody

DNA fragmentation was detected using a two-step in situ DNA fragmentation assay. Briefly, brains were fixed using 4% paraformaldehyde in 0.1 M Phosphate Buffer pH = 7.4, dehydrated and paraffin-embedded. Ten-µm parasagittal sections were rehydrated and treated with Proteinase K (1 µg/ml) for 10 min at room temperature. The DNA labelling reaction was performed at 37°C during one hour using 0.3 units/µl Terminal transferase (Boehringer Mannheim, Meylan, France) and 20 µM Biotin-16-2'-deoxyuridine-5'-triphosphate (Boehringer Mannheim). Terminal transferase was omitted in the reaction mixture used for control sections. Labelling was then detected using an avidin-biotin-peroxydase complex (Vectastain Elite ABC kit, Vector Laboratories Inc., Burlingame, CA) and DAB revelation (DAB Sigmafast, Sigma, St. Louis, MO).

Double-labelling experiments were performed on 10-µm parasagittal cryostat sections of cerebella. Sections were incubated overnight at 4°C with

CM1 antibody (dilution 1/1000), washed in PBS and used for a two-step in situ DNA fragmentation assay. The labelling reaction was performed using Biotin-1b-2'-deoxyuridine-5'-triphosphate as described above. TUNEL labelling was then revealed using Texas-red labelled streptavidin (dilution 1/1500). CM1 labelling was detected using a peroxydase-conjugated anti-rabbit antibody (dilution 1/1000) followed by FITC-conjugated tyramide deposition (NEN Life Science, Boston, MA). Control sections were included in each experiment by omitting terminal transferase or CM1 antibody or both. Results were analyzed using a Leica confocal microscope.

Histology

Animals were anesthetized using 0.1 mg/ml chloral hydrate and perfused with 0.9% sodium chloride, followed by 95% ethanol. Brains were dissected, fixed overnight in Clarke's fixative and processed for paraffin-embedding.

Ten-μm-thick parasagittal sections of the cerebellum were processed for calbindin immunohistochemistry. Sections were incubated overnight at 4 °C with CL-300 monoclonal antibody (dilution: 1/200; Sigma, St. Louis, MO). Immunocomplexes were revealed using a peroxidase-conjugated anti-mouse antibody (dilution: 1/500; Jackson ImmunoResearch, West Grove, PA) and diaminobenzidine tetrahydrochloride substrate (DAB Sigmafast, Sigma). Sections were then counterstained with cresyl violet-thioin.

Ten-μm-thick coronal sections of the brainstem were stained with cresyl-violet-thionin to locate the inferior olive and analyze its morphology. Sections from 10 different rostrocaudal levels of the inferior olive were observed in each animal to compare the morphology of all the subnuclei of the olive.

The morphology of p53 knock-out Lurcher cerebella was analyzed on sections stained with cresyl-violet-thionin.

Quantitative Analysis

The total numbers of Purkinje cells and granule cells per half-cerebellum were counted in mutant and control cerebella. Cerebellar sections were stained for calbindin immunohistochemistry and the number of calbindin-positive Purkinje cells were counted in each fortieth section at X1000 magnification. The total number of Purkinje cells was calculated from a graph of the number of Purkinje cells in each counted section plotted against the section's distance from the midline. Corrections were made for double counting Purkinje cells based on the method of Hendry (1976). We chose to use this traditional correction factor for our cell counts instead of more recently developed stereological techniques, so that our results would be directly comparable with previously published counts. The total number of granule cells per half-cerebellum was estimated from the volume of the internal granule cell layer (IGL) multiplied by the average density of granule cells in the IGL.

The granule cell density was estimated by counting, at X1000 magnification, the number of granule cells contained in an area of 25000 μm^3. These counts were done in six different regions in four sections from each half-cerebellum. Thus the numbers of granule cells in twenty-four grids were counted to obtain the average density of granule cells. The volume of the IGL was calculated from a graph of granule cell layer area plotted against distance from the midline. The area of the IGL in each fortieth section was measured using a CCD camera and NIH Image software. Corrections for double-counting errors were made using the methods of Hendry. Three to five animals of each genotype were used at each age: at P15, 3 animals of each genotype; at P30, 4 $Grid2^{Lc/+};Bax^{+/+}$, 4 $Grid2^{Lc/+};Bax^{+/-}$ and 5 $Grid2^{Lc/+};Bax^{-/-}$; at P60, 3 $Grid2^{Lc/+};Bax^{+/+}$ and 4 $Grid2^{Lc/+};Bax^{-/-}$. Statistical comparisons were made using ANOVA followed by Newman-Keuls post-hoc test (significant when $p < 0.05$).

Results

Using two different antibodies, one recognizing the two forms of caspase-3 (active and precursor) and the other one recognizing the caspase-3 precursor form in immunoblotting experiments, we have demonstrated that pro-caspase-3 expression is specifically increased in the cerebellum of $Grid2^{+/+}$ mice (data not shown; see Selimi et al. 2000c).

Localization of Caspase-3 in Wild-Type Mice

Immunohistochemistry was performed using the anti-hamster caspase-3 antibody to localize caspase-3 in the brain of $Grid2^{+/+}$ mice. In $Grid2^{+/+}$ mice, at all ages examined (P12, P15 and P20), caspase-3 immunostaining was mainly restricted to fiber-like processes (Fig. 1B). Caspase-3 immunostaining of fibers is clearly illustrated in the white matter of the cerebellum. These fibers were labelled from the white matter, through the granule cell layer to the Purkinje cell layer and were occasionally observed penetrating the molecular layer (Fig. 1B). Some of these caspase-3-immunopositive fibers were double-labelled by anti-calbindin antibodies (Fig. 1A), indicating that these fibers were Purkinje cell axons. However, not all of the caspase-3-immunopositive fibers were labelled by anti-calbindin, indicating that other categories of fibers in the cerebellum contained caspase-3. Many of the calbindin-immuno-negative fibers enveloped the soma of Purkinje cells and in some instances climbed the dendritic tree of the Purkinje cell (Fig. 1B). This morphology is characteristic of the axons of inferior olivary neurons, the cerebellar climbing fibers, indicating that this category of axons also contains caspase-3.

A polyclonal antibody directed against the p20 subunit of active caspase-3 (CM1 antibody, a kind gift of Dr. Srinivasan) was used for immunohisto-chemistry on frozen brain sections of P15 and P20 $Grid2^{+/+}$ mice to analyze

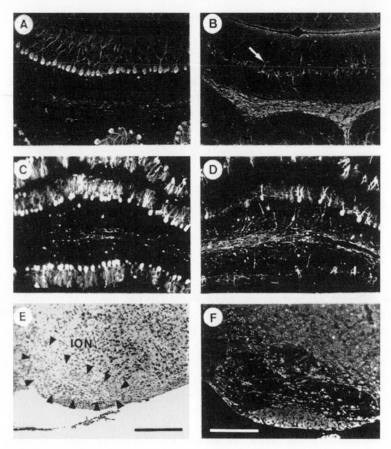

Fig. 1 A–F. Immunohistochemical localization of pro-caspase-3 in the brains of P15 Lurcher and wild-type mice. Anti-calbindin (**A, C**) and anti-caspase-3 (**B, D**) double immunostaining was performed on parasagittal sections of wild-type (**A, B**) and Lurcher (**C, D**) mouse brains. Coronal sections of the Lurcher brainstem containing the inferior olive (arrowheads) were stained with cresyl violet (**E**) and anti-caspase-3 antiserum (**F**). ION inferior olivary neuron. A–D, F, 200 µm, E, 200 µm. (From Selimi et al. 2000 b, with author's permission)

the localization of active caspase-3. In $Grid2^{+/+}$ cerebellum, Purkinje cells were consistently immunonegative and very few granule cells were labelled (Selimi et al. 2000 b). The presence of few immunopositive granule cells is consistent with the fact that developmentally regulated cell death occurs in the granule cell population during postnatal development. Occasionally, labelled neurons were also found in the inferior olive of P15 $Grid2^{+/+}$ animals (3.75 ± 1.25 labelled neurons per 10 sections). No immunostaining was found in the white matter of $Grid2^{+/+}$ cerebellum, either in the neurons of the hippocampus or the cortex of $Grid2^{+/+}$ mice.

A comparison of the immunohistochemistry results using both antibodies indicates that active caspase-3 is present only in a few granule cells of the

$Grid2^{+/+}$ cerebellum and a few neurons of the $Grid2^{+/+}$ inferior olive. Anti-hamster caspase-3 immunostaining was localized to fiber-like processes and not present in any granule cell, showing that this immunostaining in fiber-like processes is specific to pro-caspase-3.

Localization of Caspase-3 in Lurcher Mice

To identify which category of cerebellar cells overexpresses pro-caspase-3 in $Grid2^{Lc/+}$ mice, anti-hamster caspase-3 immunostaining was performed on brain sections of mutant mice.

In the cerebellum of $Grid2^{Lc/+}$ mice at all ages examined (P12, P15 and P20), caspase-3 immunostaining was strikingly intense in some, but not all, Purkinje cells (Fig. 1 D). Caspase-3 immunostaining was not detected in the cell body of any other cell type in the cerebellum (Fig. 1 D). In the cerebellar nodulus, where the Purkinje cells are the last to degenerate in the $Grid2^{Lc/+}$ mice, no caspase-3-immunopositive Purkinje cell bodies and dendrites were observed at any of the three ages examined. In coronal sections of the cerebellum and the brainstem, no staining was observed in the inferior olivary neurons (Fig. 1 F). These observations demonstrate that not all the categories of cells that die in the $Grid2^{Lc/+}$ brain are caspase-3 immunopositive. The caspase-3 immunoreactivity is specific to dying $Grid2^{Lc/+}$ Purkinje cells. In other regions of the brain not affected by the $Grid2^{Lc/+}$ mutation, no neuron was found to be immunopositive for caspase-3. For example, in the hippo-campus, where no difference of expression of caspase-3 was detected by im-munoblotting, no cell body was immunostained. These observations in $Grid2^{Lc/+}$ mice indicate a correlation between the caspase-3 immunoreactivity of Purkinje cells, the increase of pro-caspase-3 observed in the cerebellum by immunoblotting and Purkinje cell death. Caspase-3 immunostaining was also present in fiber-like processes throughout the brain of $Grid2^{Lc/+}$ animals, consistent with the result obtained from brain sections of $Grid2^{+/+}$ mice.

Interestingly, the localization of pro-caspase-3 was not the same in every immunopositive Purkinje cell. At all ages examined, numerous Purkinje cells were strongly immunoreactive in the perinuclear region of the soma and the dendritic shafts. Some Purkinje cells were stained only in the perinuclear soma, others were also stained in the major dendrites and, sometimes, cas-pase-3 staining could be observed within smaller dendrites, thus extending to the upper part of the dendritic arborization. Pro-caspase-3 was not pre-sent in the terminal parts of the dendrites, as the immunostaining did not reveal the spiny branchlets that can be seen with anti-calbindin immuno-staining (Selimi et al. 2000b). The axons of degenerating Purkinje cells in $Grid2^{Lc/+}$ mice are characterized by the presence of varicosities. These axonal varicosities were intensely stained by caspase-3 immunohistochemistry. In particular, these axonal varicosities could be strongly caspase-3 immunoposi-tive and the soma of the same Purkinje cell could be immunonegative (Fig. 1 C and D). In sections of P12 $Grid2^{Lc/+}$ mice, some Purkinje cells were stained in a more confined region of their soma. This observation suggests a confined localization of pro-caspase-3 at the beginning of the cell death process.

As we have previously noted, not all the Purkinje cells in $Grid2^{Lc/+}$ cerebellum were immunopositive. The percentage of pro-caspase-3-positive Purkinje cells was roughly the same at each age examined (20–25%), suggesting that the overexpression of pro-caspase-3 in $Grid2^{Lc/+}$ Purkinje cells is not developmentally regulated but is rather related to the degenerative state of the Purkinje cell.

To test whether caspase-3 was active in the dying $Grid2^{Lc/+}$ cells, immunohistochemistry using CM1 antibody recognizing the p20 subunit of active caspase-3 was performed on brain sections of $Grid2^{Lc/+}$ mice (P15 and P20). Strongly immunoreactive granule cells and Purkinje cells were found in $Grid2^{Lc/+}$ cerebellum (Selimi et al. 2000a). Fibers in the white matter of the cerebellum were immunonegative. No immunolabelled neurons could be found in the hippocampus and the cortex of $Grid2^{Lc/+}$ mice (which are not a priori affected by the Lc mutation). The percentage of immunopositive Purkinje cells (between 1 and 3%) was very low compared to the percentage of cells immunopositive for the caspase-3 precursor, showing that the delay between caspase-3 activation and Purkinje cell death is very short in comparison with the period of pro-caspase-3 overexpression. The number of labelled granule cells per section was considerably increased in $Grid2^{Lc/+}$ cerebellum compared to the $Grid2^{+/+}$. This number was quantified in P15 animals and the increase was found to be statistically significant. Cell counts in the inferior olive of P15 $Grid2^{Lc/+}$ mice reveal 13.86 ± 0.02 labelled neurons per 10 sections. This number is higher than the number of labelled neurons in P15 $Grid2^{+/+}$ inferior olive and corresponds to a six-fold increase in the percentage of labelled olivary neurons, when the 40% loss of $Grid2^{Lc/+}$ inferior olivary neurons at P15 is taken into account (Caddy and Biscoe 1979). The presence of active caspase-3 in granule cells and Purkinje cells of $Grid2^{Lc/+}$ cerebella was confirmed by immunohistochemical experiments using another antibody directed against active caspase-3 (MF397, a kind gift of Dr. Nicholson; data not shown). The fact that active caspase-3 could be found in granule cells and olivary neurons whereas no pro-caspase-3 was detected in these cells could be explained by a different sensitivity of the two antibodies used. For instance, the anti-hamster caspase-3 might only immunocytochemically label high levels of pro-caspase-3 in cells, such as observed in $Grid2^{Lc/+}$ Purkinje cells.

These results demonstrate that pro-caspase-3 is upregulated in dying $Grid2^{Lc/+}$ Purkinje cells. Active caspase-3 is also found in $Grid2^{Lc/+}$ Purkinje cells. In addition, active caspase-3 is present in granule cells and inferior olivary neurons that die extensively in the $Grid2^{Lc/+}$ cerebellum due to the loss of their target.

Comparison Between DNA Fragmentation and Caspase-3 Activation

Fragmentation of nuclear DNA is considered to be a general feature of cells undergoing apoptosis. To test if $Grid2^{Lc/+}$ Purkinje cells expressing caspase-3 also exhibit this characteristic, we performed an assay based on the TUNEL method (Terminal deoxynucleotidyl transferase-mediated dUTP nick-end-labelling) on brain sections of P15 $Grid2^{Lc/+}$ and $Grid^{+/+}$ mice.

To increase the sensitivity of detection, we used a two-step in situ DNA fragmentation-labelling method. Numerous granule cells were labelled in the internal granule layer of the $Grid2^{Lc/+}$ cerebellum. The number of labelled $Grid2^{Lc/+}$ granule cells was significantly higher than the number of granule cells labelled in the $Grid2^{+/+}$ cerebellum, consistent with the fact that numerous granule cells die in the $Grid2^{Lc/+}$ cerebellum following the loss of their target, the Purkinje cell. Rarely, some Purkinje cell nuclei were labelled in $Grid2^{Lc/+}$ cerebella using this two-step technique (Selimi et al. 2000b), whereas Purkinje cell nuclei were never labelled in $Grid2^{+/+}$ cerebella. We estimated the number of labelled $Grid2^{Lc/+}$ Purkinje cells to be less than 0.5%. This result suggests that DNA fragmentation in the $Grid2^{Lc/+}$ Purkinje cell is quickly followed by phagocytosis and is thus very difficult to detect.

We found that the number of TUNEL-positive granule cells and the number of active caspase-3-positive granule cells were not statistically different in $Grid2^{Lc/+}$. This suggests that the time courses of caspase-3 activation and DNA fragmentation in $Grid2^{Lc/+}$ granule cells are similar. These events are easier to detect in granule cells than in Purkinje cells, as the total number of granule cells per section is very high.

Double-labelling experiments on $Grid2^{Lc/+}$ cerebellar sections were performed using anti-active caspase-3 antibody (CM1 antibody) and a two-step in situ DNA fragmentation assay. Numerous granule cells were found to be double-labelled (Selimi et al. 2000b), confirming that granule cell death in $Grid2^{Lc/+}$ cerebellum is an apoptotic process involving caspase-3 activation and DNA fragmentation. Not all caspase-3-labelled granule cells were TUNEL-positive and vice versa. This finding indicates that these two molecular events are only partially overlapping during neuronal death, as already suggested by Namura et al. (1998). These experiments did not reveal double-labelled Purkinje cells, despite the fact that we screened sections from several animals. The number of Purkinje cells that were either active caspase-3-immunopositive or TUNEL-positive is low in $Grid2^{Lc/+}$ cerebella. Considering evidence from granule cells suggests that these two events are only partially overlapping, the probability of detecting double-labelled cells in the much smaller Purkinje cell population is remote.

Bax Deletion Doesn't Rescue the Lurcher Phenotype

Lurcher heterozygous ($Grid2^{Lc/+}$) mice were crossed with Bax knock-out ($Bax^{-/-}$) mice to determine if *Bax* inactivation rescued the Lurcher phenotype. Lurcher homozygotes were never found in the litters, suggesting that *Bax* deletion does not rescue the Lurcher homozygotes and is not sufficient to inhibit the death of brainstem neurons in these animals. The $Grid2^{Lc/+}$; $Bax^{-/-}$ mice were ataxic as the $Grid2^{Lc/+}$; $Bax^{+/+}$ controls. However, the volume of their cerebellum was greater than that of $Grid2^{Lc/+}$; $Bax^{+/+}$ mice, suggesting that *Bax* deletion does rescue some of the cells normally degenerating in Lurcher mice.

Bax Deletion Inhibits Granule Cell Death in Lurcher Mice

Purkinje cell death in $Grid2^{Lc/+}$ mice begins around P8 and this neurodegeneration is accompanied by the loss of 90% of the granule cells. By P30, 90% of Purkinje cells have already disappeared (Caddy and Biscoe 1979) and approximately 85% of granule cells (Doughty et al. 1999).

A qualitative observation of cerebellar sections taken from P30 animals showed that *Bax* deletion in $Grid2^{Lc/+}$ mice inhibited granule cell, but not Purkinje cell death. Immunohistochemistry using an anti-calbindin antibody to specifically stain Purkinje cells in the cerebellum showed that almost all Purkinje cells had degenerated in $Grid2^{Lc/+}$; $Bax^{+/+}$ controls and $Grid2^{Lc/+}$; $Bax^{-/-}$ double mutants by P30 (Selimi et al. 2000c). The morphology of $Grid2^{Lc/+}$; $Bax^{+/-}$ cerebella (data not shown) was similar to $Grid2^{Lc/+}$; $Bax^{+/+}$ cerebella. Only few Purkinje cells remained in P30 animals of all three genotypes, primarily in one lobule of the cerebellum, the nodulus. The morphology of Purkinje cells in $Grid2^{Lc/+}$; $Bax^{-/-}$ double mutants and $Grid2^{Lc/+}$; $Bax^{+/-}$ cerebella was identical to the morphology of Purkinje cells in $Grid2^{Lc/+}$; $Bax^{+/+}$ control cerebellum. The dendrites of these cells were atrophic, thicker and less branched than in normal mice as previously described (Dumesnil-Bousez and Sotelo 1992; Doughty et al. 1999). Quantitative analysis of the number of Purkunje cells per hemi-cerebellum at P30 showed that there was no significant difference between the numbers of Purkinje cells found in $Grid2^{Lc/+}$; $Bax^{+/-}$ (3208±723) and $Grid2^{Lc/+}$; $Bax^{+/+}$ (3162±298) control mice. The number of Purkinje cells in $Grid2^{Lc/+}$; $Bax^{-/-}$ double mutants (6364±520) was very low, confirming that Bax inactivation does not rescue Purkinje cells. However, this number was significantly higher than the number found in $Grid2^{Lc/+}$; $Bax^{+/-}$ and $Grid2^{Lc/+}$; $Bax^{+/+}$ mice, suggesting that Bax inactivation might delay Lurcher Purkinje cell death (Fig. 2A).

Nuclear staining of the P30 sections using cresyl violet-thionin showed a clear increase of the area of the IGL in double mutant cerebella when compared to $Grid2^{Lc/+}$; $Bax^{+/+}$ and $Grid2^{Lc/+}$; $Bax^{+/-}$ control cerebella. The number of granule cells per hemi-cerebellum was significantly higher in $Grid2^{Lc/+}$; $Bax^{-/-}$ cerebella ($10.80 \times 10^6 \pm 0.78 \times 10^6$) compared to $Grid2^{Lc/+}$; $Bax^{+/+}$ ($3.16 \times 10^6 \pm 0.35 \times 10^6$) and $Grid2^{Lc/+}$; $Bax^{+/-}$ ($2.87 \times 10^6 \pm 0.13 \times 10^6$) cerebella (Fig. 3B; ANOVA followed by post-hoc Newman-Keuls test analysis, p<0.01), confirming that *Bax* inactivation in $Grid2^{Lc/+}$, mice rescues granule cells. The

Fig. 2 A, B. Quantitative analysis of neuronal death in the cerebellum of *Bax* knock-out Lurcher mice. A, Purkinje cell number per section was assessed by using anti-calbindin-immunostained cerebellar sections taken from the vermis. B, granule cell number per section was estimated by multiplying the area of the internal granule cell layer by the average density of granule cells. Indicates that numbers found in $Grid2^{Lc/+}$; $Bax^{-/-}$ double mutant mice are statistically different from numbers found in $Grid2^{Lc/+}$; $Bax^{+/+}$ mice (p < 0.05; ANOVA followed by post-hoc Newman Keuls post-hoc test analysis). Error bars indicate SEM (n = 3–4). (Modified from Selimi et al. 2000 c, with author's permission)

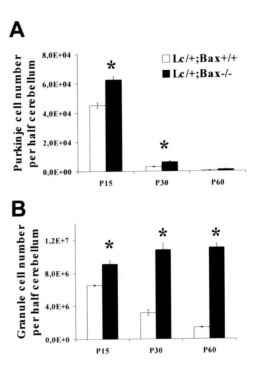

number of granule cells per hemi-cerebellum of $Grid2^{Lc/+}$; $Bax^{+/-}$ animals was not statistically different from the number of granule cells found in $Grid2^{Lc/+}$; $Bax^{+/+}$ animals, showing that inactivation of only one allele of *Bax* is not sufficient to inhibit cell death in $Grid2^{Lc/+}$ mice.

Sixty to 75% of the inferior olivary neurons normally degenerate in $Grid2^{Lc/+}$ mice following the death of their Purkinje cell targets (Caddy and Biscoe 1979; Heckroth and Eisenman 1991; Herrup et al. 1996). Fifty-four percent of these neurons have disappeared by P26 (Caddy and Biscoe 1979). To determine if *Bax* deletion was able to rescue the target-related cell death of olivary neurons in Lurcher mice, we analyzed the morphology of the inferior olive in P30 $Grid2^{Lc/+}$; $Bax^{-/-}$ double mutant animals. Coronal sections of the brainstem from 10 different rostro-caudal levels of the inferior olive were stained with cresyl violet-thionin. No obvious difference was detected between the inferior olive of $Grid2^{Lc/+}$; $Bax^{-/-}$ double mutant and $Grid2^{Lc/+}$; $Bax^{+/+}$ control mice (Selimi et al. 2000 c). A massive loss of olivary neurons was observed in all the subnuclei of the inferior olive in both $Grid2^{Lc/+}$; $Bax^{-/-}$ double mutants and $Grid2^{Lc/+}$; $Bax^{+/+}$ control mice when compared to the inferior olive of a wild-type mouse (Selimi et al. 2000 c). The rostro-caudal extension of the inferior olive was approximately 1100 μm in both $Grid2^{Lc/+}$; $Bax^{-/-}$ double mutant and $Grid2^{Lc/+}$; $Bax^{+/+}$ control mice. This value is reduced compared to the wild-type one in accordance with the results of Heckroth and Eisenman (1991). This result suggests that *Bax* inactivation

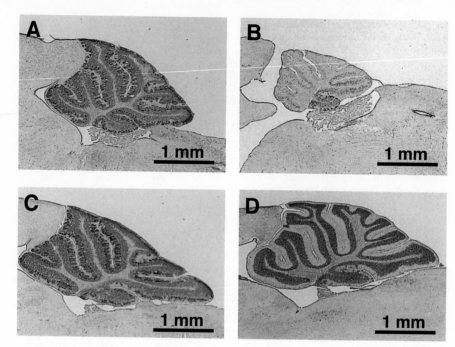

Fig. 3 A–D. Morphology of the cerebellum of control Lurcher and *Bax* knock-out Lurcher mice at P15 and two months. Purkinje cell death has already begun at P15 from $Grid2^{Lc/+};Bax^{+/+}$ controls (**A**) as well as $Grid2^{Lc/+};Bax^{-/-}$ double mutant mice (**C**), as shown by the presence of gaps in the Purkinje cell layer. The surface of the internal granule cell layer is persistently increased at two months in $Grid2^{Lc/+};Bax^{-/-}$ double mutant mice (**D**) when compared to $Grid2^{Lc/+};Bax^{+/+}$ control animals (**B**). (From Selimi et al. 2000c, with author's permission)

does not inhibit the target-related cell death of inferior olivary neurons occurring in $Grid2^{Lc/+}$ mice.

The results of the double $Grid2^{Lc/+}$ and $Bax^{-/-}$ cross at P30 show that *Bax* inactivation in Lurcher mice inhibits target-related granule cell death, but not the target-related death of inferior olivary neurons or intrinsic Purkinje cell death.

Purkinje Cell Death is Temporarily Delayed by *Bax* Inactivation in Lurcher Mice

Cerebellar sections from P15 mice were analyzed to look for an early effect of *Bax* inactivation on Lurcher Purkinje cell death. At P15, numerous Purkinje cells were still present in $Grid2^{Lc/+};Bax^{+/+}$ control and $Grid2^{Lc/+};Bax^{-/-}$ double mutant mice (Fig. 2). The presence of gaps in the Purkinje cell layer showed that Purkinje cell degeneration has already begun at this age in animals of both genotypes (Fig. 3 A and 3 C), in concordance with the observation that progressive neurodegeneration begins around P8. Quantification of the Purkinje cell number per hemi-cerebellum (Fig. 2A) showed that their number was significantly higher in $Grid2^{Lc/+}$; $Bax^{-/-}$ double mutant cerebella ($62\,680 \pm 2152$) compared to $Grid2^{Lc/+}$; $Bax^{+/+}$ cerebella ($45\,261 \pm 1813$; ANOVA followed by post-hoc Newman-Keuls test analysis, $p < 0.05$). The analysis of the latero-lateral distribution of Purkinje cells in both $Grid2^{Lc/+}$; $Bax^{-/-}$ and $Grid2^{Lc/+}$; $Bax^{+/+}$ cerebella showed that the number of Purkinje cells was increased throughout all the regions of the cerebellum in double mutants, indicating that Bax inactivation does not rescue a particular subpopulation of Purkinje cells. The number of granule cells in $Grid2^{Lc/+}$; $Bax^{-/-}$ animals was already significantly higher than in $Grid2^{Lc/+}$; $Bax^{+/+}$ mice ($9.09 \times 10^6 \pm 0.38 \times 10^6$ vs $6.50 \times 10^6 \pm 0.06 \times 10^6$, respectively; ANOVA, followed by post-hoc Newman-Keuls test analysis, $p < 0.05$) and was not significantly different from the one found in P30 animals (Fig. 2B).

These results show that Bax inactivation does not inhibit, but temporarily delays, the degeneration of some Purkinje cells in Lurcher mice until P30.

Granule Cell Rescue by *Bax* Inactivation is Persistent

The effect of Bax inactivation in P60 animals was analyzed to determine if granule cell rescue was a lasting effect in $Grid2^{Lc/+}$ double mutants (Fig. 3B and D). In $Grid2^{Lc/+}$; $Bax^{+/+}$ control mice as well as in $Grid2^{Lc/+}$; $Bax^{-/-}$ double mutant mice, Purkinje cell degeneration was almost complete at this age, except for a few cells still remaining in the nodulus (525 ± 97 vs 1268 ± 218, respectively, Fig. 3A; ANOVA followed by post-hoc Newman-Keuls test analysis, $p > 0.05$). The number of granule cells per $Grid2^{Lc/+}$; $Bax^{+/+}$ half-cerebellum was $1.30 \times 10^6 \pm 0.06 \times 10^6$, approximately 10% of the normal number of granule cells. Thus, the degeneration of granule cells is almost complete at this age in $Grid2^{Lc/+}$; $Bax^{+/+}$ control mice, as 10% of these cells still remain at P730 (Caddy and Biscoe 1979). The analysis of cerebellar sextions from P60 $Grid2^{Lc/+}$; $Bax^{-/-}$ double mutant mice (Fig. 3B) revealed that the area of the internal granule cell layer was still increased when compared to $Grid2^{Lc/+}$; $Bax^{+/+}$ control mice (Fig. 2B). The number of granule cells per hemi-cerebellum in $Grid2^{Lc/+}$; $Bax^{-/-}$ mice ($11.05 \times 10^6 \pm 0.41 \times 10^6$) was significantly higher compared to $Grid2^{Lc/+}$; $Bax^{+/+}$ controls (ANOVA followed by the Newman-Keuls post-hoc test analysis, $p < 0.05$). This number was not significantly different from the numbers of granule cells found in $Grid2^{Lc/+}$; $Bax^{-/-}$ mice at P15 and P30 (ANOVA followed by the Newman-Keuls post-hoc test analysis, $p > 0.05$). The number of rescued granule cells is approximately 70% of the number of granule cells found in wild-type animals (Vogel et al. 1991). These results indicate that granule cell rescue by Bax inactivation in Lurcher is persistent for at least two months.

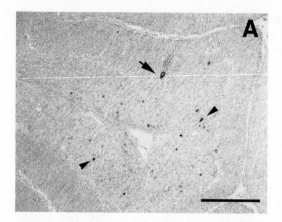

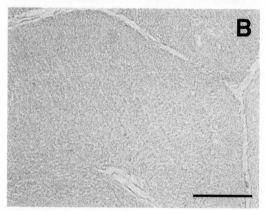

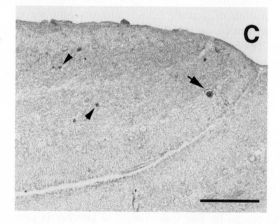

Fig. 4A–C. Effects of *Bax* and p53 inactivation on caspase-3 activation in Lurcher cerebellum. Immunohistochemical experiments were performed on 10-µm-thick parasagittal cerebellar sections using an antibody specifically recognizing activated caspase-3. In P15 control Lurcher mice (**A**) and p53 knockout Lurcher mice (**C**), activated caspase-3 is detected in Purkinje cells (arrow) and in granule cells (arrowhead). In P15 *Bax* knock-out Lurcher mice (**B**), activated caspase-3 immunoreactivity is absent, showing that *Bax* inactivation inhibits caspase-3 activation. Scale bars = 100 µm. (From Selimi et al. 2000a, with author's permission)

P53 Does not Rescue Lurcher Neuronal Death

The morphology of $Grid2^{Lc/+}$; $p53^{-/-}$ cerebellum is extremely atrophic at P30 and is imilar in appearance to the morphology of control $Grid2^{Lc/+}$ cerebellum (Selimi et al. 2000 a). Almost no Purkinje cells in $Grid2^{Lc/+}$; $p53^{-/-}$ mice are detected, except for the few remaining in the nodulus as in control $Grid2^{Lc/+}$ animals. In both mutants, the IGL is extremely reduced, showing that p53 inactivation does not inhibit $Grid2^{Lc/+}$ granule cell death. Analysis of the inferior olive in $Grid2^{Lc/+}$; $p53^{-/-}$ mice shows a neuronal loss similar to the one occuring in control $Grid2^{Lc/+}$ inferior olive (data not shown). These results show that p53 inactivation has no effect on either intrinsic Purkinje cell death or target-related neuronal death in $Grid2^{Lc/+}$ mice. Moreover, caspase-3 activation could be detected in numerous granule cells as well as in a few Purkinje cells in $Grid2^{Lc/+}$; $p53^{-/-}$ mice, showing that caspase-3 activation is not dependent on p53 in this model (Fig. 4C). Taken together, these results show that p53 expression, unlike Bax expression, is not necessary for $Grid2^{Lc/+}$ granule cell death and that both p53 inactivation and *Bax* inactivation are not able to inhibit $Grid2^{Lc/+}$ Purkinje cell death.

Effects of *Bax* and p53 Inactivation on Caspase Expression and Activation

Caspase-3 activation was analyzed in immunohistochemical experiments using an antibody specifically recognizing the activated form of caspase-3. As shown previously in P15 $Grid2^{Lc/+}$ cerebellar sections, activated caspase-3 is detected in numerous granule cells (69.6 ± 17.6 labeled granule cells per section) as well as in 1–3% Purkinje cells (Fig. 4 A; Selimi et al. 2000b). In $Grid2^{Lc/+}$; $Bax^{-/-}$ cerebella, Purkinje cells degenerate whereas approximately 70% of granule cells are rescued (Selimi et al. 2000c). Activated caspase-3 is not detected in P15 $Grid2^{Lc/+}$; $Bax^{-/-}$ cerebella, either in granule cells or in Purkinje cells, even after analysis of 15 sections per cerebellum (Fig. 4 B), showing that *Bax* is necessary for the activation of caspase-3 in dying neurons in $Grid2^{Lc/+}$ cerebella.

The effect of *Bax* and p53 inactivation on the upregulation of pro-capase-3 detected in $Grid2^{Lc/+}$ Purkinje cells was analyzed using an antibody recognizing the precursor form of pro-capase-3 (Selimi et al. 2000b). Upregulation of pro-caspase-3 expression is detected in Purkinje cells of $Grid2^{Lc/+}$; $p53^{-/-}$ and $Grid2^{Lc/+}$; $Bax^{-/-}$ mice as in control $Grid2^{Lc/+}$ mice at P15 (Fig. 1; Selimi et al. 2000a). Quantitative analysis shows that there is no statistical difference between the percentages of Purkinje cells overexpressing pro-caspase-3 in $Grid2^{Lc/+}$; $Bax^{-/-}$, $Grid2^{Lc/+}$; $p53^{-/-}$ and control $Grid2^{Lc/+}$ mice (around 35%, Kruskal-Wallis test, $p > 0.5$). Thus neither p53 nor *Bax* is involved in the up-regulation of pro-caspase-3 expression in $Grid2^{Lc/+}$ Purkinje cells.

Discussion

Studies on Lurcher chimeric mice have shown that the cerebellum of *Lc*/+ mice is affected by two types of neurodegeneration. The degeneration of Purkinje cells is a direct consequence of the presence of a mutated GRID2 protein, whereas the death of olivary neurons and granule cells is an indirect effect and presumably follows the loss of their target neuron, the Purkinje cell (Wetts and Herrup 1982 a,b; Herrup et al. 1996). Comparing both types of neuronal death in *Grid2*$^{Lc/+}$ mice suggests that different pathways are triggered depending on which death signal the neuron receives and on the neuronal population.

Caspase-3 is Constitutively Expressed in *Grid2*$^{+/+}$ Brain

A constitutive expression of caspase-3 was revealed by immunoblotting in the cortex, hippocampus and cerebellum of *Grid2*$^{+/+}$ mice, in agreement with the results obtained by other authors (Krajewska et al. 1997; Chen et al. 1998; Namura et al. 1998). We demonstrated that pro-caspase-3 is targeted to fiber-like processes in the *Grid2*$^{+/+}$ brain and we could not detect immunolabelled nerve cell bodies in agreement with Chen et al. (1998) and Krajewska et al. (1997). Namura et al. (1998) could detect pro-caspase-3 in fiber-like processes and in some neuronal somas. These observations suggest that, if pro-caspase-3 is present in neuronal somas, its level is quite low and not easily detectable by immunohistochemistry. In our preparation, some cerebellar granule cells of the *Grid2*$^{+/+}$ mice contained active caspase-3, which is likely due to developmentally regulated cell death (Herrup and Busser 1995). Taken together, these results indicate that some categories of neurons express the apoptotic machinery even in the normal postnatal brain.

The Intrinsic *Grid2*$^{Lc/+}$ Purkinje Cell Death

All but a few cerebellar Purkinje cells, 90% of the granule cells and 60 to 75% of the olivary neurons degenerate postnatally in the *Grid2*$^{Lc/+}$ mutant (Phillips 1960; Caddy and Biscoe 1979). The genetic defect responsible for Purkinje cell degeneration in *Grid2*$^{Lc/+}$ mice is a point mutation in the gene encoding the GRID2 protein. This protein has an unknown function but the *Lc* mutation in this protein leads to the presence of a constitutive inward current in the *Grid2*$^{Lc/+}$ Purkinje cells that maintains these cells at a depolarized resting membrane potential. These observations have led to the idea that the *Grid2*$^{Lc/+}$ Purkinje cell death is induced by excitotoxicity (Zuo et al. 1997). However, the molecular mechanism underlying this neuronal death has not been resolved. *Grid2*$^{Lc/+}$ Purkinje cell death has been described as necrotic based on morphological criteria (Dumesnil-Bousez and Sotelo 1992).

We have shown that caspase-3 is upregulated and activated in dying Lurcher ($Grid2^{Lc/+}$) Purkinje cells. Pro-caspase-3 was detected in the perinuclear soma as well as in the dendritic shafts of Purkinje cells. Its subcellular localization was shown to be cytoplasmic and partly mitochondrial. DNA fragmentation was found using the TUNEL technique in the nuclei of a few Purkinje cells in the cerebellum of $Grid2^{Lc/+}$ mice. These observations suggest that the death of Purkinje cells in $Grid2^{Lc/+}$ mice is mediated by caspase-3 activity.

$Grid2^{Lc/+}$ Purkinje cell death might involve two phases: the first phase involves an increase in the expression of pro-caspase-3, followed by a subsequent triggering of caspase-3 activation engaging the Purkinje cell in the last phase of its destruction. Mancini et al. (1998) have shown that the fraction of pro-caspase-3 present in mitochondria disappears after the induction of apoptosis. Pro-caspase-3 was detected in the mitochondria of dying $Grid2^{Lc/+}$ Purkinje cells overexpressing pro-caspase-3, indicating that these cells might be in the first phase of their death process when apoptosis is not yet fully triggered. The first morphological signs of $Grid2^{Lc/+}$ Purkinje cell abnormalities can be detected as early as P6–P8 (Dumesnil-Bousez and Sotelo 1992; Bailly et al. 1996). Interestingly, one of these signs is a swelling of the mitochondria, a feature that has been observed in some models of apoptosis (Vander Heiden et al. 1997). The death signal in $Grid2^{Lc/+}$ Purkinje cell might also induce a mitochondrial dysfunction and the subsequent release, at an adequate time, of mitochondrial components that have been shown to induce caspase-3 activation and the subsequent destruction of the cell, e.g., cytochrome c or AIF (Green and Reed 1998). Finally, both pathways, mitochondrial dysfunction and caspase activation, might interact to amplify each other, as has recently been proposed by Green and Kroemer (1998), leading to a rapid final destruction of the $Grid2^{Lc/+}$ Purkinje cells. In fact, the actual destruction of the cell, the second phase of $Grid2^{Lc/+}$ Purkinje cell death, might be triggered only once a certain threshold of mitochondrial dysfunction and death protein overexpression is reached. This destruction ends by a rapid engulfment of the remnants of the cell (Thornberry and Lazebnik 1998). Thus, the second phase of the $Grid2^{Lc/+}$ Purkinje cell death process is likely to be more difficult to observe, a hypothesis that is supported by the fact that, in the $Grid2^{Lc/+}$ cerebellum, the number of Purkinje cells containing active caspase-3 is quite low, and the number of TUNEL-positive Purkinje cells is even lower.

Zanjani et al. (1998 b) demonstrated that the overexpression of the anti-apoptotic protein BCL-2 does not rescue the $Grid2^{Lc/+}$ Purkinje cells. Thus, the intrinsic neuronal death affecting $Grid2^{Lc/+}$ Purkinje cells is BCL-2-independent and mediated by the up-regulation and activation of caspase-3. During the first phase, an upregulation of caspase-3 expression is induced. It is interesting to note that such an up-regulation has already been revealed in another model of excitotoxicity, ischemia (Chen et al. 1998; Namura et al. 1998). In addition, caspase-3 overexpression is able to induce apoptosis in vitro (Fernandes-Alnemri et al. 1994). These results suggest that the upregulation of pro-caspases is also involved in the induction of neuronal apoptosis in vivo, especially in the case of a chronic exposure to a death signal such as occurs in $Grid2^{Lc/+}$ Purkinje cells. Although BCL-2 overexpression does not inhibit $Grid2^{Lc/+}$ Purkinje cell death, it is able to delay the death process, as, unlike in control $Grid2^{Lc/+}$ mice, Purkinje cells can be found in two-month-old $Grid2^{Lc/+}$ mice overexpressing BCL-2 (Zanjani et al. 1998 a). Interestingly, BCL-2 has been shown to downregulate caspase-3 expression in neuronal precursors (Korhonen et al. 1997). Therefore, BCL-2 might downregulate temporarily the overexpression of caspase-3 in $Grid2^{Lc/+}$ Purkinje cells and might also inhibit the effect of BAX, a pro-apoptotic protein, that has recently been shown to be upregulated in dying $Grid2^{Lc/+}$ Purkinje cells (Wullner et al. 1998). But, as the abnormal inward current induced by the Lurcher mutation of GRID2 in $Grid2^{Lc/+}$ Purkinje cells is continually present (Zuo et al. 1997), BCL-2 overexpression might become insufficient.

One of the mechanisms by which BCL-2 may exert its neuroprotective action is through inhibition of the pro-apoptotic protein, BAX (Oltvai et al. 1993; Antonsson et al. 1997; Mahajan et al. 1998). BAX mRNA is expressed at high levels in the postnatal mouse brain (de Bilbao et al. 1999). BAX protein expression levels may gradually decline in the cerebellum through the first 21 days of development, although expression levels of BAX remain high in Purkinje cells in the adult (Vekrellis et al. 1997). BAX plays an important role in naturally occurring cell death since neuronal cell death is reduced in $Bax^{-/-}$ mice; for example, the number of facial nucleus motoneurons is increased by 51% (Deckwerth et al. 1996; White et al. 1998).

BAX expression is increased in dying $Grid2^{Lc/+}$ PCs, suggesting that it may play a role in their death (Wullner et al. 1998). Yet, it is surprising that deletion of BAX expression only slightly delays $Grid2^{Lc/+}$ Purkinje cell death. The number of $Grid2^{Lc/+}$ Purkinje cells is significantly increased at P15 in the $Grid2^{Lc/+}$ double mutant, but by P30 this number has returned to a low level and at P60 is not significantly different from the one found in the $Grid2^{Lc/+}$ mutant. The available evidence suggests that $Grid2^{Lc/+}$ PCs may die by an excitotoxic mechanism induced by the accumulation of leaky GRID2 subunits at developing PC-parallel fiber synapses (Zuo et al. 1997; De Jager and Heintz 1998). The Lurcher gain-of-function mutation in $Grid2$ results in a large, constitutive inward sodium current in the cells that express the subunit and might induce a defect in Ca^{2+} homeostasis (Zuo et al. 1997). While we cannot rule out a role for BAX-induced apoptosis in $Grid2^{Lc/+}$ Purkinje cells, their death in Bax knock-out Lurcher mutants indicates that BAX is not necessary for their demise. One likely hypothesis is that there are multiple pathways and/or other BCL-2 family members capable of inducing cell death in stressed cells, including Purkinje cells. The slight delay in $Grid2^{Lc/+}$ PC death seen at P15 may represent the difference between the BAX cell death usually induced in $Grid2^{Lc/+}$ mutants and a different mechanism induced in $Grid2^{Lc/+}; Bax^{-/-}$ double mutants. BCL-2 overexpression may be able to temporarily rescue $Grid2^{Lc/+}$ Purkinje cells (Zanjani et al. 1998 a) due to its ability to bind to BAX and prevent the formation of BAX channels (Antonsson et al. 1997; Mahajan et al. 1998) and/or its ability to regulate intra-cellular Ca^{2+} homeostatis (Murphy et al. 1996; Ichimiya et al. 1998; Kuo et al. 1998). But in the end, the $Grid2^{Lc/+}$ mutation is lethal and neither *Bax* inacti-vation-or *bcl-2* overexpression can rescue Lurcher Purkinje cells.

We have shown in vivo that pro-caspase-3 up-regulation is not depen-dent on BAX, in concordance with the in vitro results of Miller et al. (1997) who showed that the increase in caspase-3 mRNA levels observed in K^+-de-prived granule cells is not dependent on BAX. Moreover, this finding sug-gests that the delay in $Grid2^{Lc/+}$ Purkinje cell death induced by *Bax* inactiva-tion is not a consequence of the inhibition of pro-caspase-3 upregulation. Our results also show that pro-caspase-3 upregulation is not dependent on p53 and suegest that the gene encoding pro-caspase-3 might be upregulated by another pathway. For example, the family of STAT genes (Signal Transdu-cer and Activator of Transcription) has been suggested to regulate caspase expression (Fu 1999). p53 inactivation has been shown to inhibit excitotoxic neuronal death in several models (Hughes et al. 1997). However, our results show that p53 inactivation does not inhibit caspase-3 activation and Pur-kinje cell death in $Grid2^{Lc/+}$ mice. Unlike p53, *Bax* inactivation inhibits cas-pase-3 activation in $Grid2^{Lc/+}$ Purkinje cells and delays their death. The delay induced by *Bax* inactivation could be due to inhibition of the caspase activa-tion cascade. However, *Bax* inactivation is not sufficient to prevent Purkinje cell death. Indeed, *Bax* and caspase-3 inactivation also only rescue transi-ently the neuronal death induced by β-amyloid peptide in an in vitro model (Giovanni et al. 2000). Our results suggest that multiple pathways may lead to the death induced by a permanent depolarization of the Purkinje cells. Other proteins of the BCL-2 family could be involved in this death process. Alternatively, the involvement of another protease, calpain, which can be di-rectly activated by an increase in Ca^{2+} concentration, has been demonstrated in models of excitotoxicity (Wang 2000). Thus, this calpain-dependent path-way would not be inhibited in $Grid2^{Lc/+};Bax^{-/-}$ Purkinje cells and could be responsible for the death of these neurons. Our results suggest that in some neurodegenerative diseases the inhibition of a single death pathway would be an insufficient treatment to prevent neuronal death.

Target-Related Cell Death in $Grid2^{Lc/+}$ Cerebellum

In the olivocerebellar system, surgical and genetic lesions have shown that the survival of olivary neurons and granule cells is dependent on interactions with their Purkinje cell targets (Sotelo and Changeux 1974; Caddy and Biscoe 1979; Wetts and Herrup 1982 a,b; Zanjani et al. 1990; Vogel et al. 1991; Herrup et al. 1996). Many olivary neurons and granule cells undergo apoptotic cell death when deprived of their target (Smeyne et al. 1995; Chu et al. 2000).

In the $Grid2^{Lc/+}$ mutant, numerous granule cells contain the active cas-pase-3 and are TUNEL-positive in $Grid2^{Lc/+}$ cerebellum. Some are double-labelled, confirming the apoptotic nature of granule cell death. Target-related cell death of granule cells has already been associated with TUNEL staining in the Lurcher cerebellum (Wüllner et al. 1995) and also in a model where Purkinje cells were ablated by the specific expression of diphtheria toxin (Smeyne et al. 1995). The presence of active caspase-3 is also detected in some olivary neurons in $Grid2^{Lc/+}$ brainstem. In parallel, pro-caspase-3 is not upregulated during the target-related death of the granule cells and olivary neurons in $Grid2^{Lc/+}$ mice. In contrast to Purkinje cells, the overexpression of BCL-2 in Lurcher does rescue inferior olivary neurons (Zanjani et al. 1998b). Bcl-2 was not overexpressed in the granule cells in this transgenic model of Bcl-2 overexpression. However, granule cells from these mice overexpress BCL-2 in primary cultures and are resistant to various death stimuli (Tanabe et al. 1997). Thus, on the basis that granule cells and inferior olivary neurons are subjected to similar cell death stimuli, we can speculate that granule cells would also be rescued by BCL-2 overexpression in vivo. Taken together, these observations suggest that the target-related cell death of granule cells and olivary neurons in $Grid2^{Lc/+}$ mice involves an apoptotic pathway that is BCL-2-dependent and does not require pro-caspase-3 up-regulation but is mediated by caspase-3 activation.

There are multiple lines of evidence that support a role for BCL-2-related proteins, including BAX, in target-related cell death. We have previously shown that overexpression of bcl-2 in $Grid2^{Lc/+}$ olivary neurons will rescue most of the olivary neurons from target-related cell death (Zanjani et al. 1998b). This result is consistent with previous studies showing in vitro rescue of neurons from trophic factor withdrawal (Allsopp et al. 1993; Batista-tou et al. 1993; Garcia et al. 1992) and in vivo rescue of neurons from axotomy or ischemia by bcl-2 overexpression (Dubois-Dauphin et al. 1994; Martinou et al. 1994; Farlie et al. 1995; Bonfanti et al. 1996; de Bilbao and Dubois-Dauphin 1996) or deletion of Bax expression (Deckwerth et al. 1996; White et al. 1998).

However, our results reveal differences in the pathway leading to this target-related cell death in vivo: deletion of Bax expression in $Grid2^{Lc/+}$ mutants rescues granule cells but does not prevent olivary neuron death. Both populations are dying due to loss of their target, but granule cell death appears to be dependent on Bax expression whereas olivary neuron cell death can be independent of Bax. Granule cell rescue could have been a consequence of the delay in Purkinje cell death. For example, in pcd mutant mice, Purkinje cell death occurs between the third and sixth postnatal week and the number of granule cells is not different from controls in P30 animals. However, this number is significantly decreased in three-month-old pcd mutants and follows an exponential decay, suggesting that a transient trophic support from Purkinje cells is not sufficient to inhibit persistently target-related granule cell death (Triarhou 1998). Not even a slight decrease in granule cell numbers was observed in Bax knock-out Lurcher mice from P15 to P60, suggesting that the delay in Purkinje cell death is unlikely to be the only cause of granule cell rescue. Our estimates of granule cell numbers indicate that there is an eight-fold increase in the number of surviving granule cells in the $Grid2^{Lc/+}; Bax^{-/-}$ double mutants compared to $Grid2^{Lc/+}; Bax^{+/+}$ controls. Granule cell numbers in our Bax knock-out Lurcher mutants are only reduced by 30% compared to wild-type numbers (Vogel et al. 1991). The number of granule cells in $Grid2^{Lc/+}; Bax^{-/-}$ double mutants is comparable to the maximum number of granule cells detected in $Grid2^{Lc/+}; Bax^{+/+}$ cerebella (8×10^6; Caddy and Biscoe 1979). Thus, Bax inactivation does not significantly affect the deficit in granule cell genesis detected in Lurcher mutants. It seems more likely that the primary cause of granule cell rescue in Bax knock-out Lurcher mice is the loss of BAX expression. Wild type and Lurcher granule cells have been shown to express Bax during the period of degeneration in Lurcher cerebellum (Vekrellis et al. 1997; Wullner et al. 1998) and our results suggest that Bax plays an important role in regulating granule cell death following removal of target-related trophic support.

Bax inactivation in $Grid2^{Lc/+}$ mice is able to inhibit target-related granule cell death until at least P300 (Doughty et al. 2000; Selimi et al. 2000c). Our results confirm that Bax inactivation inhibits caspase-3 activation in $Grid2^{Lc/+}$ Purkinje cells and granule cells. BAX has been shown previously to facilitate cytochrome c release from the mitochondria, thereby activating the effector caspases and apoptosis (Desagher et al. 1999; Gross et al. 1999). Thus, our results suggest that the inhibition of caspase-3 activation in $Grid2^{Lc/+}; Bax^{-/-}$ mice might be one reason for the inhibition of granule cell death in this model. However, we cannot exclude other effects of Bax inactivation that could play a role in preventing granule cell death. Indeed, in an in vitro model of granule cell death, BAX has been shown to initiate a caspase-dependent as well as a caspase-independent death process (Miller et al. 1997). The fact that Bax inactivation inhibits at least one downstream event participating to apoptosis, that is caspase-3 activation, strengthens the idea that granule cell rescue in this model is a direct effect of the inhibition of apoptosis. This finding suggests that the temporary rescue of some Purkinje cells between P15 and P30, and the trophic support they might supply, is not responsible for granule cell rescue by Bax inactivation.

p53 is able to induce BAX-dependent caspase-3 activation and apoptosis in granule cells in vitro (Cregan et al. 1999). In contrast, we show that p53 inactivation does not inhibit caspase-3 activation and apoptosis in granule cells of $Grid2^{Lc/+}$ mice. Taken together, our results suggest that granule cell death in $Grid2^{Lc/+}$ mice is mediated by a p53-independent, BAX-dependent caspase-3 activation.

The molecular pathways that link trophic factor deprivation with cell death have not yet been fully characterized. The binding of trophic factors with their receptors triggers a variety of signaling responses that promote cell survival, in some cases through interactions with BCL-2-related proteins (Kaplan and Miller 1997; Pettmann and Henderson 1998). For example, NGF or IGF-1 can induce the phosphorylation of BAD, a non-membrane bound BCL-2 relative, by activating Akt, a serine-threonine protein kinase. In the absence of trophic factor, non-phosphorylated BAD will bind to BCL-x and prevent BCL-x's anti-apoptotic activity. BCL-x's anti-apoptotic activity may involve binding to BAX to prevent it from opening a pore in mitochondrial membranes (Gross et al. 1999). Recently, nerve growth factor (NGF) has also been shown to promote BCL-2 expression in NGF-responsive neurons (Riccio et al. 1999). It is not yet clear what pathway or pathways are involved in cell death in granule cells and olivary neurons. There is a large variety of BCL-2-related cell death proteins that promote or block cell death (Gross et al. 1999). For example, the loss of *Bax* expression in *Bcl-x*-deficient mice prevents most, but not all, cell death, so there must be other cell death-promoting proteins (Schindler et al. 1997). The rescue of granule cells, but not olivary neurons, by *Bax* inactivation following target loss indicates that while BAX is required for target-related cell death in granule cells, there may be other cell death-promoting proteins functioning in olivary neurons.

References

Adams JM, Cory S (1998) The Bcl-2 protein family: arbiters of cell survival. Science 281:1322–1326

Allsopp TE, Wyatt S, Paterson HF, Davies AM (1993) The proto-oncogene bcl-2 can selectively rescue neurotrophic factor-dependent neurons from apoptosis. Cell 73:295–307

Antonsson B, Conti F, Ciavatta A, Montessuit S, Lewis S, Martinou I, Bernasconi L, Bernard A, Mermod JJ, Mazzei G, Maundrell K, Gambale F, Sadoul R, Martinou JC (1997) Inhibition of Bax channel-forming activity by Bcl-2. Science 277:370–372

Araki K, Meguro H, Kushiya E, Takayama C, Inoue Y, Mishina M (1993) Selective expression of the glutamate receptor channel delta 2 subunit in cerebellar Purkinje cells. Biochem Biophys Res Commun 197:1267–1276

Bailly Y, Kyriakopoulou K, Delhaye-Bouchaud N, Mariani J, Karagogeos D (1996) Cerebellar granule cell differentiation in mutant and X-irradiated rodents revealed by the neural adhesion molecule TAG-1. J Comp Neurol 369:150–161

Batistatou A, Merry DE, Korsmeyer SJ,, Greene LA (1993) Bcl-2 affects survival but not neuronal differentiation of PC12 cells. J Neurosci 13(10):4422–4428

Bonfanti L, Strettoi E, Chierzi S, Cenni MC, Liu XH, Martinou JC, Maffei L, Rabacchi SA (1996) Protection of retinal ganglion cells from natural and axotomy-induced cell death in neonatal transgenic mice overexpressing bcl-2. J Neurosci 16:4186–4194

Caddy KW, Biscoe TJ (1979) Structural and quantitative studies on the normal C3H and Lurcher mutant mouse. Phil Trans R Soc Lond B Biol Sci 287:167–201

Chen J, Nagayama T, Jin K, Stetler RA, Zhu RL, Graham SH, Simon RP (1998) Induction of caspase-3-like protease may mediate delayed neuronal death in the hippocampus after transient cerebral ischemia. J Neurosci 18:4914–4928

Chu T, Hullinger H, Schilling K, Oberdick J (2000) Spatial and temporal changes in natural and target deprivation-induced cell death in the mouse inferior olive. J Neurobiol 43:18–30

Cregan SP, MacLaurin JG, Craig CG, Robertson GS, Nicholson DW, Park DS, Slack RS (1999) Bax-dependent caspase-3 activation is a key determinant in p53-induced apoptosis in neurons. J Neursci 19:7860–7869

de Bilbao F, Dubois-Dauphin M (1996) Time course of axotomy-induced apoptotic cell death in facial motoneurons of neonatal wild type and bcl-2 transgenic mice. Neuroscience 71:1111–1119

de Bilbao F, Guarin E, Nef P, Vallet P, Giannakopoulos P, Dubois-Dauphin M (1999) Postnatal distribution of cpp32/caspase-3 mRNA in the mouse central nervous system: an in situ hybridization study. J Comp Neurol 409:339–357

Deckwerth TL, Elliott JL, Knudson CM, Johnson EM Jr, Snider WD, Korsmeyer SJ (1996) BAX is required for neuronal death after trophic factor deprivation and during development. Neuron 17:401–411

De Jager PL, Heintz N (1998) The lurcher mutation and ionotropic glutamate receptors: contributions to programmed neuronal death in vivo. Brain Pathol 8:795–807

Desagher S, Osen-Sand A, Nichols A, Eskes R, Montessuit S, Lauper S, Maundrell K, Antonsson B, Martinou JC (1999) Bid-induced conformational change of Bax is responsible for mitochondrial cytochrome c release during apoptosis. J Cell Biol 144:891–901

Doughty ML, Lohof A, Selimi F, Delhaye-Bouchaud N, Mariani J (1999) Afferent-target cell interactions in the cerebellum: negative effect of granule cells on Purkinje cell development in lurcher mice. J Neurosci 19:3448–3456

Doughty ML, De Jager PL, Korsmeyer SJ, Heintz N (2000) Neurodegeneration in lurcher mice occurs via multiple cell death pathways [In Process Citation]. J Neurosci 20:3687–3694

Du Y, Dodel RC, Bales KR, Jemmerson R, Hamilton-Byrd E, Paul SM (1997) Involvement of a caspase-3-like cysteine protease in 1-methyl-4-phenylpyridinium-mediated apoptosis of cultured cerebellar granule neurons. J Neurochem 69:1382–1388

Dubois-Dauphin M, Frankowski H, Tsujimoto Y, Huarte Y, Martinou J-C (1994) Neonatal motoneurons overexpressing the bcl-2 proto-oncogene in transgenic mice are protected from axotomy-induced cell death. Proc Natl Acad Sci USA 91:3309–3313

Dumesnil-Bousez N, Sotelo C (1992) Early development of the Lurcher cerebellum: Purkinje cell alterations and impairment of synaptogenesis. J Neurocytol 21:506–529

Farlie PG, Dringen R, Rees SM, Kannourakis G, Bernard O (1995) bcl-2 transgene expression can protect neurons against developmental and induced cell death. Proc Natl Acad Sci USA 92:4397–4401

Fernandes-Alnemri T, Litwack G, Alnemri ES (1994) CPP32, a novel human apoptotic protein with homology to *Caenorhabditis elegans* cell death protein Ced-3 and mammalian interleukin-1β-converting enzyme. J Biol Chem 269:30761–30764

Fu XY (1999) From PTK-STAT signaling to caspase expression and apoptosis induction [see comments]. Cell Death Differ 6:1201–1208

Garcia I, Martinou I, Tsujimoto Y, Martinou J (1992) Prevention of programmed cell death of sympathetic neurons by the *bcl*-2 proto-oncogene. Science 258:302–304

Giovanni A, Keramaris E, Morris EJ, Hou ST, O'Hare M, Dyson N, Robertson GS, Slack RS, Park DS (2000) E2F1 mediates death of β-amyloid-treated cortical neurons in a manner independent of p53 and dependent on bax and caspase-3. J Biol Chem 275:11553–11560

Green DR, Kroemer G (1998) The central executioners of apoptosis: caspases or mitochondria? Trends Cell Biol 8:267–271

Green DR, Reed JC (1998) Mitochondria and apoptosis. Science 281(5381):1309–1312

Gross A, McDonnell JM, Korsmeyer SJ (1999) BCL-2 family members and the mitochondria in apoptosis. Genes Dev 13:1899–1911

Heckroth JA, Eisenman LM (1991) Olivary morphology and olivocerebellar topography in adult lurcher mutant mice. J Comp Neurol 312:641–651

Hendry IA (1976) A method to correct adequately for the change in neuronal size when estimating neuronal numbers after growth factor treatment. J Neurocytol 5:337–349

Herrup K, Busser JC (1995) The induction of multiple cell cycle events precedes target-related neuronal death. Development 121:2385–2395

Herrup K, Shojaeian-Zanjani H, Panzini L, Sunter K, Mariani J (1996) The numerical matching of source and target populations in the CNS: the inferior olive to Purkinje cell projection. Brain Res Dev Brain Res 96:28–35

Hughes PE, Alexi T, Schreiber SS (1997) A role for the tumour suppressor gene p53 in regulating neuronal apoptosis. Neuroreport 8:V–XII

Ichimiya M, Chang SH, Liu H, Berezesky IK, Trump BF, Amstad PA (1998) Effect of Bcl-2 on oxidant-induced cell death and intracellular Ca^{2+} mobilization. Am J Physiol 275(3 Pt 1):C832–C839

Kaplan DR, Miller FD (1997) Signal transduction by the neurotrophin receptors. Curr Opin Cell Biol 9:213–221

Knudson CM, Tung KS, Tourtellotte WG, Brown GA, Korsmeyer SJ (1995) Bax-deficient mice with lymphoid hyperplasia and male germ cell death. Science 270:96–99

Kohda K, Wang Y, Yuzaki M (2000) Mutation of a glutamate receptor motif reveals its role in gating and delta2 receptor channel properties [see comments]. Nat Neurosci 3:315–322

Krajewska M, Wang HG, Krajewski S, Zapata JM, Shabaik A, Gascoyne R, Reed JC (1997) Immunohistochemical analysis of in vivo patterns of expression of CPP32 (Caspase-3), a cell death protease. Cancer Res 57:1605–1613

Kuida K, Zheng TS, Na S, Kuan C, Yang D, Karasuyama H, Rakic P, Flavell RA (1996) Decreased apoptosis in the brain and premature lethality in CPP32-deficient mice. Nature 384:368–372

Kuo TH, Kim HR, Zhu L, Yu Y, Lin HM, Tsang W (1998) Modulation of endoplasmic reticulum calcium pump by Bcl-2. Oncogene 17:1903–1910

Mahajan NP, Linder K, Berry G, Gordon GW, Heim R, Herman B (1998) Bcl-2 and Bax interactions in mitochondria probed with green fluorescent protein and fluorescence resonance energy transfer [see comments]. Nat Biotechnol 16:547–552

Mancini M, Nicholson DW, Roy S, Thornberry NA, Peterson EP, Casciola-Rosen LA, Rosen A (1998) The caspase-3 precursor has a cytosolic and mitochondrial distribution: implications for apoptotic signaling. J Cell Biol 140:1485–1495

Marks N, Berg MJ, Guidotti A, Saito M (1998) Activation of caspase-3 and apoptosis in cerebellar granule cells. J Neurosci Res 52:334–341

Martinou JC, Dubois-Dauphin M, Staple JK, Rodriguez I, Frankowski H, Missotten M, Albertini P, Talabot D, Catsicas S, Pietra C, Huarte J (1994) Overexpression of BCL-2 in transgenic mice protects neurons from naturally occurring cell death and experimental ischemia. Neuron 13:1017–1030

Miller TM, Moulder KL, Knudson CM, Creedon DJ, Deshmukh M, Korsmeyer SJ, Johnson EM Jr (1997) Bax deletion further orders the cell death pathway in cerebellar granule cells and suggests a caspase-independent pathway to cell death. J Cell Biol 139:205–217

Murphy AN, Bredesen DE, Cortopassi G, Wang E, Fiskum G (1996) Bcl-2 potentiates the maximal calcium uptake capacity of neural cell mitochondria. Proc Natl Acad Sci USA 93:9893–9898

Namura S, Zhu J, Fink K, Endres M, Srinivasan A, Tomaselli KJ, Yuan J, Moskowitz MA (1998) Activation and cleavage of caspase-3 in apoptosis induced by experimental cerebral ischemia. J Neurosci 18(10):3659–3668

Nicholson DW, Thornberry NA (1997) Caspases: killer proteases. Trends Biochem Sci 22:299–306

Norman DJ, Feng L, Cheng SS, Gubbay J, Chan E, Heintz N (1995) The lurcher gene induces apoptotic death in cerebellar Purkinje cells. Development 121:1183–1193

Oltvai ZN, Milliman CL, Korsmeyer SJ (1993) Bcl-2 heterodimerizes in vivo with a conserved homolog, Bax, that accelerates programmed cell death. Cell 74:609–619

Pettmann B, Henderson CE (1998) Neuronal cell death. Neuron 20:633–647

Phillips RJS (1960) Lurcher, a new gene in linkage group XI of the house mouse. J Genet 57:35–42

Riccio A, Ahn S, Davenport CM, Blendy JA, Ginty DD (1999) Mediation by a CREB family transcription factor of NGF-dependent survival of sympathetic neurons. Science 286:2358–2361

Selimi F, Campana A, Weitzman J, Vogel M, Mariani J (2000a) Bax and p53 are differentially involved in the regulation of caspase-3 expression and activation during neurodegeneration in Lurcher mice. CR Acad Sci III 323:1–7

Selimi F, Doughty M, Delhaye-Bouchaud N, Mariani J (2000b) Target-related and intrinsic neuronal death in Lurcher mutant mice are both mediated by caspase-3 activation. J Neurosci 20:992–1000

Selimi F, Vogel MW, Mariani J (2000c) Bax inactivation in lurcher mutants rescues cerebellar granule cells but not purkinje cells or inferior olivary neurons. J Neurosci 20:5339–5345

Shindler KS, Latham CB, Roth KA (1997) Bax deficiency prevents the increased cell death of immature neurons in bcl-x-deficient mice. J Neurosci 17:3112–3119

Smeyne R, Chu T, Lewin A, Bian F, S-Crisman S, Kunsch C, Lira S, Oberdick J (1995) Local control of granule cell generation by cerebellar Purkinje cells. Mol Cell Neurosci 6:230–251

Sotelo C, Changeux JP (1974) Transsynaptic degeneration 'en cascade' in the cerebellar cortex of staggerer mutant mice. Brain Res 67:519–526

Tanabe H, Eguchi Y, Kamada S, Martinou J, Tsujimoto Y (1997) Susceptibility of cerebellar granule neurons derived from Bcl-2-deficient and transgenic mice to cell death. Eur J Neurosci 9:848–856

Thornberry NA, Lazebnik Y (1998) Caspases: enemies within. Science 281:1312–1316

Thornberry NA, Rano TA, Peterson EP, Rasper DM, Timkey T, Garcia-Calvo M, Houtzager VM, Nordstrom PA, Roy S, Vaillancourt JP, Chapman KT, Nicholson DW (1997) A combinatorial approach defines specificities of members of the caspase family and granzyme B. J Biol Chem 272:17907–17911

Triarhou LC (1998) Rate of neuronal fallout in a transsynaptic cerebellar model. Brain Res Bull 47:219–222

Vander Heiden MG, Chandel NS, Williamson EK, Schumacker PT, Thompson CB (1997) Bcl-x$_L$ regulates the membrane potential and volume homeostasis of mitochondria. Cell 91:627–637

Vekrellis K, McCarthy MJ, Watson A, Whitfield J, Rubin LL, Ham J (1997) Bax promotes neuronal cell death and is downregulated during the development of the nervous system. Development 124:1239–1249

Vogel MW, McInnes M, Zanjani HS, Herrup K (1991) Cerebellar Purkinje cells provide target support over a limited spatial range: evidence from lurcher chimeric mice. Brain Res Dev Brain Res 64:87–94

Wetts R, Herrup K (1982a) Interaction of granule, Purkinje and inferior olivary neurons in lurcher chimaeric mice. I. Qualitative studies. J Embryol Exp Morphol 68:87–98

Wetts R, Herrup K (1982b) Interaction of granule, Purkinje and inferior olivary neurons in lurcher chimeric mice. II. Granule cell death. Brain Res 250:358–362

White FA, Keller-Peck CR, Knudson CM, Korsmeyer SJ, Snider WD (1998) Widespread elimination of naturally occurring neuronal death in Bax-deficient mice. J Neurosci 18:1428–1439

Wullner U, Loschmann PA, Weller M, Klockgether T (1995) Apoptotic cell death in the cerebellum of mutant weaver and lurcher mice. Neurosci Lett 200:109–112

Wullner U, Weller M, Schulz JB, Krajewski S, Reed JC, Klockgether T (1998) Bcl-2, Bax and Bcl-x expression in neuronal apoptosis: a study of mutant weaver and lurcher mice. Acta Neropathol (Berl) 96:233–238

Zanjani HS, Mariani J, Herrup K (1990) Cell loss in the inferior olive of the staggerer mutant mouse is an indirect effect of the gene. J Neurogenet 6:229–241

Zanjani HS, Rondi-Reig L, Vogel M, Martinou JC, Delhaye-Bouchaud N, Mariani J (1998a) Overexpression of a Hu-bcl-2 transgene in Lurcher mutant mice delays Purkinje cell death. CR Acad Sci III 321:633–640

Zanjani HS, Vogel MW, Martinou JC, Delhaye-Bouchaud N, Mariani J (1998b) Postnatal expression of Hu-bcl-2 gene in Lurcher mutant mice fails to rescue Purkinje cells but protects inferior olivary neurons from target-related cell death. J Neurosci 18:319–327

Zuo J, De Jager PL, Takahashi KA, Jiang W, Linden DJ, Heintz N (1997) Neurodegeneration in Lurcher mice caused by mutation in delta2 glutamate receptor gene. Nature 388:769–773

Neuronal Death in Huntington's Disease: Multiple Pathways for One Issue?

S. Humbert and F. Saudou

Summary

Huntington's disease (HD) is a mid-life onset neurodegenerative disorder characterized by involuntary movements (chorea), personality changes and dementia. The neuropathology of HD is a marked neuronal death in the striatum while other brain structures are selectively spared. The defective gene in HD contains a trinucleotide CAG repeat expansion within its coding region that is expressed as a polyglutamine repeat in the protein huntingtin. CAG expansions represent a novel type of mutation in the human genome and have also been found in several other inherited neurodegenerative disorders. The mechanisms by which mutant huntingtin induces neuronal death are not well understood. However, studies suggest a cascade of events that begins in the cytoplasm and ends with the translocation and accumulation of the mutant protein in the nucleus. During this course, mutant huntingtin might affect a wide range of intracellular systems such as vesicular transport and trafficking and the apoptotic machinery as well as transcription. Modification of one if not all of those intracellular systems could ultimately lead to the death of the striatal neurons in HD.

Introduction

Huntington's disease (HD) is a devastating, inherited neurodegenerative disease characterized by chorea, personality changes, dementia and an early death (Martin and Gusella 1986). The characteristic symptoms of patients with HD result from the selective death and dysfunction of specific neuronal subpopulations within the central nervous system. HD leads to significant neuronal death within the striatum, the subcortical brain structure that controls body movements, and to a lesser extent within the cortex (Vonsattel et al. 1985). The specificity of neuronal death seen in HD is striking: within the striatum, the enkephalin-containing medium spiny neurons of the striatum are particularly vulnerable, whereas many of the neighboring neurons remain unaffected (Graveland et al. 1985; Reiner et al. 1988; Albin et al. 1992; Hedreen and Folstein 1995; Sapp et al. 1995).

Linkage analysis of families with HD led to the identification of the gene IT-15, whose mutation causes HD (Group 1993). The wild-type gene encodes

Henderson/Green/Mariani/Christen (Eds.)
Neuronal Death by Accident or by Design
© Springer-Verlag Berlin Heidelberg 2001

a 350 kD protein, huntingtin, that bears no homology to known proteins. The huntingtin gene contains a polymorphic stretch of repeated CAG trinucleotides that encodes a polyglutamine tract within huntingtin. When the number of repeats exceeds 35, the gene encodes a version of huntingtin that leads to disease. Although normal huntingtin function remains an enigma, studies of mice in which the huntingtin gene has been disrupted and then restored have shown that huntingtin is required for normal embryonic development and neurogenesis (Duyao et al. 1995; Nasir et al. 1995; Zeitlin et al. 1995; White et al. 1997).

Since the absence of huntingtin leads to embryonic lethality, a phenotype that is distinct from mid-life neurodegeneration seen in HD, huntingtin mutations that lead to HD are unlikely to cause disease simply by producing a loss of huntingtin function. Furthermore, the fact that the lack of huntingtin does not recapitulate HD symptoms and that HD is inherited dominantly and affects heterozygotes and homozygotes similarly have led to the idea that mutated huntingtin gains a new toxic function (Gusella and MacDonald 1995; Sharp and Ross 1996). The expansion of the polyglutamine tract within huntingtin may create a protein epitope that is distinct from epitopes found in the wild-type huntingtin protein (Trottier et al. 1995b). The new epitope has been hypothesized to change the interactions that occur between huntingtin and other proteins, thereby leading to neurodegeneration. Proteins that interact with huntingtin could play a role in huntingtin function in health and disease (Li et al. 1995; Bao et al. 1996; Burke et al. 1996; Kalchman et al. 1996, 1997; Liu et al. 1997; Faber et al. 1998; Gusella and MacDonald 1998; Sittler et al. 1998).

Identification of the gene responsible for HD has undoubtedly allowed a better understanding of the mechanisms responsible for the disease. However, the exact molecular events leading to the death of the striatal neurons are still not well understood. For example, the huntingtin protein is expressed widely throughout the central nervous system and in non-neuronal cells yet only a small subset of these cells (i.e., the medium spiny neurons of the striatum) dies in HD (Gutekunst et al. 1995; Sharp et al. 1995; Trottier et al. 1995a). Understanding the mechanisms that regulate huntingtin-induced toxicity in striatal neurons may reveal a variety of signal transduction pathways and, hopefully, will lead to therapeutic interventions for patients.

The Cytoplasm as the Starting Point

As a first step towards understanding huntingtin's mechanism of action and the basis for its neuronal subtype specificity, efforts have been aimed at defining the subcellular site of huntingtin action. In several studies huntingtin has been found within perikarya, neurites and at synapses (DiFiglia et al. 1995; Gutekunst et al. 1995; Sharp et al. 1995; Trottier et al. 1995a; Saudou et al. 1996). Although wild type huntingtin was also described in the nucleus (De Rooij et al. 1996), the main localization of wild type huntingtin is in the cytoplasm. Huntingtin has been shown to associate with vesicle membranes and microtubules in vitro and in neurons (DiFiglia et al. 1995; Tukamoto et al. 1997). Huntingtin, as part of a protein complex that includes HAP-1 and/ or HIP-1, two huntingtin-associated proteins, may participate in endocytosis and in the microtubule-dependent transport of organelles (Block-Galarza et al. 1997; Gutekunst et al. 1998; Li et al. 1998; Velier et al. 1998; Engqvist-Goldstein et al. 1999; Kim et al. 1999b; Martin et al. 1999; Kegel et al. 2000).

As will be described later, several lines of evidence suggest that the main site of the toxic action of huntingtin is within the nucleus. However, studies also suggest that polyglutamine expansion in huntingtin might have some deleterious consequences in the cytoplasm. Analysis of a YAC transgenic mouse model reveals early electrophysiological abnormalities that occur before neurodegeneration, indicating a cytoplasmic dysfunction prior to cell death (Hodgson et al. 1999). Furthermore, a study of a mouse model of HD in which the wild type allele was replaced by a mutant allele containing 70 to 80 glutamine repeats (Shelbourne et al. 1999) showed cytoplasmic alterations, as revealed by the presence of neuropil aggregates in axons and in axon terminals (Li et al. 2000). In vitro, striatal neurons transfected by mutant huntingtin showed hallmarks of neuritic degeneration before intranuclear inclusions were observed and before cell death occured (Li et al. 2000). Finally, analysis of post mortem brains from HD patients revealed the presence of dystrophic neurites in early stages of HD (Sapp et al. 1999). Taken together, those observations suggest that, in HD, early alterations such as dysfunction of neuritic transport in the cytoplasm and/or in the neuronal processes may represent the first pathological events that could trigger a cascade that ultimately leads to the death of the striatal neurons (Li 2000). This hypothesis is supported by the recent observation that HIP-1, a huntingtin-interacting protein, possesses some proapoptotic properties (Hackam et al. 2001). HIP-1 is a cytoplasmic protein that has some homology with the yeast cytoskeletal assembly gene SLA2p (Kalchman et al. 1997; Wanker et al. 1997). HIP-1 contains a novel death effector domain, activates caspases such as caspase 3 and is able to induce apoptosis in transfected cells (Hackam et al. 2000). HIP-1 has a stronger interaction with the wild type huntingtin compared to the mutant huntingtin containing the expanded polyglutamine stretch (Kalchman et al. 1997). In the normal situation HIP-1 binds to the wild type huntingtin and is not able to induce apoptosis. In contrast, when HIP-1 is co-expressed with the mutant protein, HIP-1 does not associate with it and is then able to activate the apoptotic machinery (Hackam et al. 2000). HIP-1 activation could either represent one of the first events leading to the activation of the apoptotic machinery in the cytoplasm or could be an independent pathway that could also participate in the HD pathogenesis.

Huntingtin Processing and Translocation

Huntingtin is mainly found in the cytoplasm. However, in the pathological situation huntingtin aggregates in the nucleus as neuronal intranuclear inclusions (see Fig. 1 and below). Interestingly, only *N*-terminal fragments of huntingtin are found in the nucleus, suggesting a cellular process leading to the generation of one or more *N*-terminal fragments of huntingtin that contain the polyglutamine stretch and that translocate into the nucleus. The observation that *N*-terminal huntingtin fragments cause disease in mouse transgenic models (Mangiarini et al. 1996; Schilling et al. 1999) has suggested that abnormal polyglutamine expansions might render huntingtin toxic by facilitating its proteolysis into more toxic *N*-terminal fragments (Martindale et al. 1998). The molecular events that lead to the generation of *N*-terminal frag-

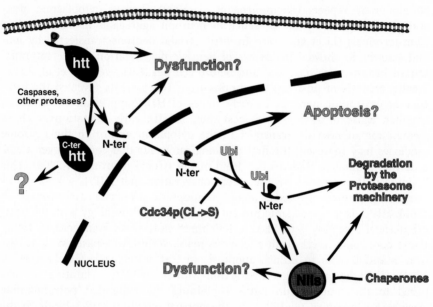

Fig. 1. Mutant huntingtin might activate multiple pathways to induce dysfunction and death of striatal neurons. Mutant huntingtin is localized in the cytoplasm. Cleavage by caspases leads to the generation of *N*-terminal fragment(s) that translocate into the nucleus. In the cytoplasm, full-length huntingtin or the truncated fragment could lead to dysfunction and/or activate the apoptotic machinery. When localized in the nucleus, mutant huntingtin induces apoptosis by unknown mechanisms that could involve alteration of the transcriptional machinery or specific protein interactions. Activation of the apoptotic machinery either by the presence of the huntingtin in the cytoplasm or in the nucleus might lead by a loop mechanism to an increased cleavage of the full-length huntingtin. In the nucleus, mutant huntingtin aggregates into intranuclear inclusions. These intranuclear inclusions are not directly responsible for the induction of apoptosis but they could modify, for example, the transcriptional machinery and thereby play a role in the induction of neuronal dysfunction. Intranuclear inclusion formation is a reversible process that is regulated by the ubiquitin-proteasome machinery as well as by chaperone proteins. Htt, huntingtin; N-ter, *N*-terminal fragment(s) of huntingtin; Ubi, ubiquitin; NIIs, neuronal intranuclear inclusions

ments may involve several caspases as the responsible proteases. Indeed, huntingtin is cleaved in vitro by caspases 1, 3 and 6 (Goldberg et al. 1996; Wellington et al. 1998). Mutations of the caspase 3 and 6 cleavage sites in polyglutamine-containing mutant huntingtin lead to a reduction, in transfected cells, of huntingtin-induced cell death and aggregate formation (Wellington et al. 2000). These results suggest that caspase 3 and 6 are involved in the generation of an N-terminal fragment with higher toxicity (Wellington et al. 2000). Huntingtin proteolysis has been demonstrated in HD patients (DiFiglia et al. 1997). However, it is unknown which proteases cleave huntingtin in vivo, and the direct involvement of caspase 3 and 6 has not yet been established in HD patients. It is interesting to note that if huntingtin needs to be proteolyzed to become toxic, the selective neuronal loss in HD might arise from cell-specific expression of huntingtin proteases (Goldberg et al. 1996).

The Nucleus as a Necessary Checkpoint?

Studies of transgenic mouse models of HD and of the brains of HD patients have shown that mutant huntingtin can form aggregates within the nucleus, suggesting that mutant huntingtin might induce neurodegeneration by acting in the nucleus (Davies et al. 1997; DiFiglia et al. 1997; Becker et al. 1998).

To characterize the mechanisms of huntingtin-induced neurodegeneration, we have developed a cellular model of HD by introducing forms of wild-type or mutant huntingtin into cultured striatal neurons by transfection and then assaying the effect of huntingtin expression on neuronal survival (Saudou et al. 1998). We found that transfected huntingtin is capable of inducing neurodegeneration in a cell-specific manner that depends on the presence of the mutated polyglutamine stretch. We also found that mutant huntingtin forms intranuclear aggregates that stain positively for huntingtin and ubiquitin. The intranuclear inclusions appear similar to the intranuclear inclusions described in HD patients and in HD transgenic mice (Davies et al. 1997; DiFiglia et al. 1997; Becher et al. 1998). Since in this in vitro cellular model mutant but not wild type huntingtin accumulates within the nucleus, we tested whether nuclear localization was necessary for huntingtin-induced neurodegeneration. To block the nuclear localization of huntingtin, we added a small nuclear export signal (NES) onto the N-terminus of the first 171 amino acids of the huntingtin protein. If huntingtin acts within the nucleus to trigger apoptosis, then the addition of an NES to huntingtin should block the ability of huntingtin to enter the nucleus and to induce neurodegeneration. On the other hand, if huntingtin acts in the cytoplasm to induce neuronal degeneration, then the addition of an NES to huntingtin might block the accumulation of huntingtin within the nucleus but it should increase apoptosis. Whereas mutant huntingtin lacking an NES led to the formation of intranuclear inclusions within approximately 30% of striatal neurons at nine days following transfection, the addition of an NES completely blocked inclusion formation within neurons transfected by mutant huntingtin-NES (Saudou et al. 1998). In these conditions, we found that, when mutant huntingtin is exported from the nucleus, the huntingtin protein completely loses its ability to induce striatal neurodegeneration (Saudou et al. 1998). Taken together, these results suggest that mutant huntingtin is transported from its site of synthesis in the cytoplasm into the nucleus where it acts by unknown mechanisms to trigger apoptosis. The fact that in HD the mutant protein acts within the nucleus to induce cell death may be generalized to the other polyglutamine neurodegenerative disorders. Indeed, results obtained in vivo by Orr and collaborators using transgenic mouse models of spinocerebellar ataxia-1 (SCA1) demonstrated that nuclear localization of mutant ataxin-1 is necessary to induce the pathology in mice (Klement et al. 1998).

Nuclear Inclusions and Toxicity

Intranuclear inclusions are associated with the pathology of HD and the other polyglutamine neurodegenerative disorders, and the inclusions form before neurological symptoms and neurodegeneration occur, leading some to suggest that the aggregation of huntingtin into intranuclear inclusions is a required step in neurodegeneration (Ross 1997; Davies et al. 1998). However, in our study (Saudou et al. 1998), we found several instances in which the formation of intranuclear inclusions did not correlate with death, suggesting the possibility that inclusions per se are not sufficient to mediate huntingtin toxicity. For example, hippocampal neurons transfected with mutant forms of huntingtin exhibit more inclusions than similarly transfected striatal neurons but hippocampal neurons do not subsequently undergo apoptosis. Also, we have identified several agents that protect against huntingtin toxicity that also increase the fraction of neurons that exhibit intranuclear inclusions. Together these data suggest that intranuclear inclusions are not toxic themselves. This hypothesis is further supported by a study also showing a dissociation between intranuclear inclusions and cell death in another in vitro model of HD (Kim et al. 1999a). In addition, in post mortem brains from HD patients, the distribution of intranuclear inclusions does not correlate with the degenerating regions (Gutekunst et al. 1999; Kuemmerle et al. 1999).

The lack of association between intranuclear inclusions and pathology could be generalized to the other polyglutamine neurodegenerative disorders, such as SCA1. Indeed transgenic models showed that a form of ataxin-1 containing the polyglutamine stretch but also a mutation in its self-association domain does not aggregate into intranuclear inclusions but is able to induce a progressive neurological phenotype (Klement et al. 1998).

In HD, intranuclear inclusions are composed of mutant huntingtin and ubiquitin (Davies et al. 1997). We have hypothesized that manipulating the ubiquitin proteasome machinery could affect the process of huntingtin aggregation. To test this idea, we coexpressed with mutant huntingtin a dominant-interfering form of ubiquitin conjugating enzyme (hCdc34p(CL->S)). We found that expression of this dominant interfering form drastically suppressed the formation of intranuclear inclusions. However, in this situation we found that mutant huntingtin-induced cell death was accelerated (Saudou et al. 1998), suggesting that ubiquitination could be a central step in the regulation of intranuclear inclusion formation and toxicity. Furthermore these results showing a strong dissociation between aggregation and cell death suggest that accumulation of mutant huntingtin into intranuclear inclusions may represent a phenomenon associated with a protective mechanism in the neurons to degrade the toxic mutant huntingtin. The involvement of the ubiquitin-proteasome machinery in the degradation of the toxic proteins in polyglutamine neurodegenerative disorders is supported by the fact that inhibition of the ubiquitination process in ataxin-1 mice results in a decrease in the formation of intranuclear inclusions but an increased progression of the disease phenotype (Cummings et al. 1999).

The ubiquitin-proteasome machinery may represent a central mechanism in the formation of intranuclear inclusions (Fig. 1). However, other proteins or factors may also participate in the regulation of intranuclear inclusion formation. The most effective proteins to regulate intranuclear inclusions are the heat shock proteins that participate in the unfolding of misfolded proteins (Chai et al. 1999; Stenoien et al. 1999; Carmichael et al. 2000; Kazemi-Esfarjani and Benzer 2000; Kobayashi et al. 2000; Muchowski et al. 2000; Satyal et al. 2000; Wyttenbach et al. 2000). In addition, intranuclear inclusion formation is reversible in a conditional transgenic model of HD, suggesting that intranuclear inclusion are dynamic structures whose formation is highly regulated (Yamamoto et al. 2000). If intranuclear inclusions are not toxic themselves, they could still play a role in HD pathogenesis. Indeed, transgenic mice carrying the exon 1 of huntingtin show a massive presence of intranuclear inclusions and develop motor symptoms (Mangiarini et al. 1996; Davies et al. 1997). However, in these mice, apoptotic cell death has never been demonstrated (Turmaine et al. 2000). This fact suggests that the symptoms observed in mice may result from the presence of intranuclear inclusions and neuronal dysfunction rather than cell death. Identification of new factors that regulate the formation of intranuclear inclusions may reveal additional ways to regulate mutant huntingtin-induced neuronal dysfunction and toxicity.

Huntingtin and Transcription

Polyglutamine-rich regions can act in vitro as activation domains for transcription factors (Gerber et al. 1994). Moreover, large numbers of proteins that contain a polyglutamine stretch are themselves transcription factors or transcriptional activators. These include proteins such as TATA-binding protein (TBP) or CREB-binding protein (CBP; Gerber et al. 1994; and, for a review, see Cha 2000). Given the fact that mutant huntingtin contains an aberrant polyglutamine stretch and acts within the nucleus to induce cell death, it has been proposed that mutant huntingtin might affect transcription. Recent studies support this hypothesis.

Mutant huntingtin has been shown to recruit transcription factors such as TBP into inclusions in the post mortem brains of HD patients (Huang et al. 1998). In vitro, CBP is recruited into aggregates, raising the possibility that CREB-mediated transcriptional activity might be altered in HD patients (Kazantsev et al. 1999). This hypothesis is supported by the fact that exon 1 of huntingtin containing the polyglutamine stretch interacts in vitro with CBP and that, in a transgenic mouse model of HD, CBP localizes in intranuclear inclusions (Steffan et al. 2000). The decrease in CREB-mediated transcription might be general to the polyglutamine neurodegenerative disorders. CBP is also recruited into intranuclear inclusions, both in vitro and in post mortem brains from spinal and bulbar muscular atrophy (SBMA) patients (McCampbell et al. 2000). This recruitment of CBP leads to a reduction of the soluble CBP and CBP-dependent transcription. Another way of interfering with CREB-mediated transcription could involve an interaction between $TAF_{II}130$ and stretches of polyglutamines (Shimohata et al. 2000). $TAF_{II}130$ colocalizes with atrophin-1 and ataxin-3 in post mortem brains from patients affected by the Dentato-Rubro-Pallidoluysian Atrophy (DRPLA) and the SpinoCerebellar Ataxia type 3 (SCA3), respectively. Using truncated forms of the atrophin-1, Tsuji and colleagues have shown that polyglutamine stretches interact with $TAF_{II}130$ (but not with CREB or CBP) and suppress CREB-dependent transcriptional activation (Shimohata et al. 2000).

If an alteration of CREB-dependent transcription is a likely mechanism, polyglutamine stretches could also modulate transcription by other pathways. Steffan and collaborators (2000) have shown that the exon 1 of the huntingtin protein containing the polyglutamine stretch interacts with p53 and represses transcription of p53-regulated promoters. Repression of transcription could also be mediated by the interaction between mutant huntingtin and N-CoR, a nuclear receptor co-repressor (Boutell et al. 1999). In this situation repression could occur through the formation of a complex that also include mSin3A, a cofactor of transcription and histone deacetylases. Similar complexes involving histone deacetylase activities might also modulate the toxicity of ataxin-1, since Sin3 has the capacity to enhance the mutant ataxin-1-induced neurodegeneration in a Drosophila model of SCA1 (Fernandez-Funez et al. 2000).

Taken together, these studies suggest that polyglutamine-containing proteins will alter the transcriptional machinery when they are localized in the nucleus. These studies also suggest that a wide range of transcription factors might be involved in this process. Furthermore, oligonucleotide DNA arrays profiling experiments show that about 1% of the tested transcripts have an altered expression in a transgenic model of HD (Luthi-Carter et al. 2000).

This finding raises the possibility that the transcriptional alteration specifically induced by the mutant protein might modify the rate of transcription of numerous genes, rending difficult the identification of the key gene(s) involved in the death process.

Huntingtin and Apoptosis

Increased apoptosis has been detected in post mortem brains from HD patients, suggesting that mutant huntingtin induces neurodegeneration by activating the apoptotic machinery (Dragunow et al. 1995; Portera-Cailliau et al. 1995; Thomas et al. 1995). Apoptosis is also detected in in vitro and in vivo models fo HD (Jackson et al. 1998; Reddy et al. 1998; Saudou et al. 1998; Warrick et al. 1998; Faber et al. 1999; Kim et al. 1999a; Miyashita et al. 1999; Wang et al. 1999; Rigamonti et al. 2000). Using our in vitro model of huntingtin-induced neurodegeneration, we found that mutant huntingtin induces cell death by an apoptotic mechanism and that mutant huntingtin-induced cell death is blocked by the antiapoptotic protein Bcl-xL as well as by Ac-DEVD-CHO, an inhibitor of caspase 3 (Saudou et al. 1998). In this assay we used a fragment of huntingtin that corresponds to the first 480 amino acids of the mouse protein that includes the polyglutamine stretch. This fragment was chosen because it corresponds to the N-terminal fragment obtained when huntingtin is cleaved by CPP32 (Goldberg et al. 1996). The fact that the neuronal death induced by this N-terminal fragment of huntingtin containing the polyglutamine stretch is blocked by caspase 3 inhibitor suggests that, in addition to cleaving the full length protein, caspase 3 may also have a role downstream of the generation of N-terminal fragments of huntingtin. Finally, it is tempting to speculate that, by a loop mechanism, increased activation of caspase-3 might, in addition to inducing the cleavage of a wide range of substrates, lead to an increased accumulation of the toxic N-terminal fragment of huntingtin. This mechanism might also take place in other polyglutamine neurodegenerative disorders, since proteins such as atrophin-1 or the androgen receptor are also cleaved by caspases and in some cases the cleavage of those proteins has been shown to be important for mediating the toxicity (Miyashita et al. 1997; Wellington et al. 1998; Ellerby et al. 1999a,b). Other caspases than caspase 3 could also play a role in huntingtin-induced neuronal death. In transgenic mice models of HD, inhibition of caspase 1 results in the delay of the progression of the pathology (Ona et al. 1999; Chen et al. 2000). Furthermore caspase 1 could also be involved in the generation of more toxic N-terminal fragments containing the polyglutamine stretch (Wellington et al. 1998).

Many studies suggest the involvement of several components of the apoptotic machinery in HD (Saudou et al. 1998; Miyashita et al. 1999; Sanchez et al. 1999; Hackam et al. 2000; Rigamonti et al. 2000). How polyglutamine stretches activate these components is not clear. Several signaling pathways that could lead to polyglutamine-dependent activation of apoptosis have been proposed (Liu 1998; Sawa et al. 1999; Hackam et al. 2001; Kouroku et al. 2000; Liu et al. 2000). In addition, alteration of the transcriptional machinery might also deregulate the cellular homeostasis, thereby leading to the activation of apoptosis.

In conclusion, a review of the recent literature suggests that expansion of glutamines in mutant huntingtin leads to the gain of a toxic function for the protein. When mutated, huntingtin may alter several pathways, which in turn will induce the death of striatal neurons in the brain of HD patients. Understanding the new toxic function of the mutant huntingtin is clearly a priority in HD research. However, in patients, the normal function of huntingtin may also be modified since conditional knock-out of the huntingtin in the adult mouse lead to neurodegeneration (Dragatsis et al. 2000). There is currently not treatment to delay or prevent the appearance of symptoms in patients. The recent studies suggest that inhibitors of apoptosis are valuable candidates as therapeutic agents in HD patients since they could inhibit in vivo both the cleavage of huntingtin into a more toxic N-terminal fragment and the subsequent death induced by this fragment (Wellington and Hayden 2000). However, activation of apoptosis in the striatal neurons could represent a final event, occurring after neuronal dysfunction and the appearance of the symptoms in HD patients. Apoptosis could then be activated to selectively eliminate the non-functional neurons. In this case, anti-apoptotic compounds might be beneficial for neuronal survival but disastrous for the proper function of the neurons and the neuronal circuitry involved in HD. A better understanding of all the mechanisms that participate in HD pathology will hopefully help in the design of new drugs that could prevent both neuronal dysfunction and death.

Acknowledgments. The authors thank Ela Bryson and Hélène Rangone for critical reading of the manuscript. Sandrine Humbert is supported by a fellowship from Association pour la Recherche contre le Cancer (ARC, France). Frédéric Saudou is an investigator from Institut National de la Santé et de la Recherche Médicale (INSERM, France) and is supported by fellowships from the Centre National de la Recherche Scientifique (ATIPE, CNRS, France), Association pour la Recherche contre le Cancer (ARC, France) and the CHDI from the Hereditary Disease Foundation.

References

Albin RL, Reiner A, Anderson KD, Dure LSt, Handelin B, Balfour R, Whetsell WO Jr, Penney JB, Young AB (1992) Preferential loss of striato-external pallidal projection neurons in presymptomatic Huntington's disease. Ann Neurol 31:425–430

Bao J, Sharp AH, Wagster MV, Becher M, Schilling G, Ross CA, Dawson VL, Dawson TM (1996) Expansion of polyglutamine repeat in huntingtin leads to abnormal protein interactions involving calmodulin. Proc Natl Acad Sci USA 93:5037–5042

Becher MW, Kotzuk JA, Sharp AH, Davies SW, Bates GP, Price DL, Ross CA (1998) Intranuclear neuronal inclusions in Huntington's disease and dentatorubral and pallidoluysian atrophy: correlation between the density of inclusions and IT15 CAG triplet repeat length. Neurobiol Dis 4:387–397

Block-Galarza J, Chase KO, Sapp E, Vaughn KT, Vallee RB, DiFiglia M, Aronin N (1997) Fast transport and retrograde movement of huntingtin and HAP 1 in axons. Neuroreport 8:2247–2251

Boutell JM, Thomas P, Neal JW, Weston VJ, Duce J, Harper PS, Jones AL (1999) Aberrant interactions of transcriptional repressor proteins with the Huntington's disease gene product, huntingtin. Human Mol Genet 8:1647–1655

Burke JR, Enghild JJ, Martin ME, Jou YS, Myers RM, Roses AD, Vance JM, Strittmatter WJ (1996) Huntingtin and DRPLA proteins selectively interact with the enzyme GAPDH. Nat Med 2:347–350

Carmichael J, Chatellier J, Woolfson A, Milstein C, Fersht AR, Rubinsztein DC (2000) Bacterial and yeast chaperones reduce both aggregate formation and cell death in mammalian cell models of Huntington's disease. Proc Natl Acad Sci USA 97:9701–9705

Cha JH (2000) Transcriptional dysregulation in Huntington's disease. Trends Neurosci 23:387–392

Chai Y, Koppenhafer SL, Bonini NM, Paulson HL (1999) Analysis of the role of heat shock protein (Hsp) molecular chaperones in polyglutamine disease. J Neurosci 19:10338–10347

Chen M, Ona VO, Li M, Ferrante RJ, Fink KB, Zhu S, Bian J, Guo L, Farrell LA, Hersch SM, Hobbs W, Vonsattel JP, Cha JH, Friedlander RM (2000) Minocycline inhibits caspase-1 and caspase-3 expression and delays mortality in a transgenic mouse model of Huntington disease. Nat Med 6:797–801

Cummings CJ, Reinstein E, Sun Y, Antalffy B, Jiang Y, Ciechanover A, Orr HT, Beaudet AL, Zoghbi HY (1999) Mutation of the E6-AP ubiquitin ligase reduces nuclear inclusion frequency while accelerating polyglutamine-induced pathology in SCA1 mice. Neuron 24:879–892

Davies SW, Turmaine M, Cozens BA, DiFiglia M, Sharp AH, Ross CA, Scherzinger E, Wanker EE, Mangiarini L, Bates GP (1997) Formation of neuronal intranuclear inclusions underlies the neurological dysfunction in mice transgenic for the HD mutation. Cell 90:537–548

Davies SW, Beardsall K, Turmaine M, DiFiglia M, Aronin N, Bates GP (1998) Are neuronal intranuclear inclusions the common neuropathology of triplet-repeat disorders with polyglutamine-repeat expansions? Lancet 351:131–133

De Rooij KE, Dorsman JC, Smoor MA, Den Dunnen JT, Van Ommen GJ (1996) Subcellular localization of the Huntington's disease gene product in cell lines by immunofluorescence and biochemical subcellular fractionation. Human Mol Genet 5:1093–1099

DiFiglia M, Sapp E, Chase K, Schwarz C, Meloni A, Young C, Martin E, Vonsattel JP, Carraway R, Reeves SA, Boyce F, Aronin N (1995) Huntingtin is a cytoplasmic protein associated with vesicles in human and rat brain neurons. Neuron 14:1075–1081

DiFiglia M, Sapp E, Chase KO, Davies SW, Bates GP, Vonsattel JP, Aronin N (1997) Aggregation of huntingtin in neuronal intranuclear inclusions and dystrophic neurites in brain. Science 277:1990–1993

Dragatsis I, Levine MS, Zeitlin S (2000) Inactivation of hdh in the brain and testis results in progressive neurodegeneration and sterility in mice. Nat Genet 26:300–306

Dragunow M, Faull RL, Lawlor P, Beilharz EJ, Singleton K, Walker EB, Mee E (1995) In situ evidence for DNA fragmentation in Huntington's disease striatum and Alzheimer's disease temporal lobes. Neuroreport 6:1053–1057

Duyao MP, Auerbach AB, Ryan A, Persichetti F, Barnes GT, McNeil SM, Ge P, Vonsattel JP, Gusella JF, Joyner AL, MacDonald ME (1995) Inactivation of the mouse Huntington's disease gene homolog Hdh. Science 269:407–410

Ellerby LM, Andrusiak RL, Wellington CL, Hackam AS, Propp SS, Wood JD, Sharp AH, Margolis RL, Ross CA, Salvesen GS, Hayden MR, Bredesen DE (1999a) Cleavage of atrophin-1 at caspase site aspartic acid 109 modulates cytotoxicity. J Biol Chem 274:8730–8736

Ellerby LM, Hackam AS, Propp SS, Ellerby HM, Rabizadeh S, Cashman NR, Trifiro MA, Pinsky L, Wellington CL, Salvesen GS, Hayden MR, Bredesen DE (1999b) Kennedy's disease: caspase cleavage of the androgen receptor is a crucial event in cytotoxicity. J Neurochem 72:185–195

Engqvist-Goldstein AE, Kessels MM, Chopra VS, Hayden MR, Drubin DG (1999) An actin-binding protein of the Sla2/Huntingtin interacting protein 1 family is a novel component of clathrin-coated pits and vesicles. J Cell Biol 147:1503–1518

Faber PW, Barnes GT, Srinidhi J, Chen J, Gusella JF, MacDonald ME (1998) Huntingtin interacts with a family of WW domain proteins. Human Mol Genet 7:1463–1474

Faber PW, Alter JR, MacDonald ME, Hart AC (1999) Polyglutamine-mediated dysfunction and apoptotic death of a Caenorhabditis elegans sensory neuron. Proc Natl Acad Sci USA 96:179–184

Fernandez-Funez P, Nino-Rosales ML, de Gouyon B, She WC, Luchak JM, Martinez P, Turiegano E, Benito J, Capovilla M, Skinner PJ, McCall A, Canal I, Orr HT, Zoghbi HY, Botas J (2000) Identification of genes that modify ataxin-1-induced neurodegeneration. Nature 408:101–106

Gerber HP, Seipel K, Georgiev O, Hofferer M, Hug M, Rusconi S, Schaffner W (1994) Transcriptional activation modulated by homopolymeric glutamine and proline stretches. Science 263:808–811

Goldberg YP, Nicholson DW, Rasper DM, Kalchman MA, Koide HB, Graham RK, Bromm M, Kazemi-Esfarjani P, Thornberry NA, Vaillancourt JP, Hayden MR (1996) Cleavage of huntingtin by apopain, a proapoptotic cysteine protease, is modulated by the polyglutamine tract. Nat Genet 13:442–449

Graveland GA, Williams RS, DiFiglia M (1985) Evidence for degenerative and regenerative changes in neostriatal spiny neurons in Huntington's disease. Science 227:770–773

Group THsDCR (1993) A novel gene containing a trinucleotide repeat that is expanded and unstable on Huntington's disease chromosomes. Cell 72:971–983

Gusella JF, MacDonald ME (1995) Huntington's disease. Sem Cell Biol 6:21–28

Gusella JF, MacDonald ME (1998) Huntingtin: a single bait hooks many species. Curr Opin Neurobiol 8:425–430

Gutekunst CA, Levey AI, Heilman CJ, Whaley WL, Yi H, Nash NR, Rees HD, Madden JJ, Hersch SM (1995) Identification and localization of huntingtin in brain and human lymphoblastoid cell lines with anti-fusion protein antibodies. Proc Natl Acad Sci USA 92:8710–8714

Gutekunst CA, Li SH, Yi H, Ferrante RJ, Li XJ, Hersch SM (1998) The cellular and subcellular localization of huntingtin-associated protein 1 (HAP1): comparison with huntingtin in rat and human. J Neurosci 18:7674–7686

Gutekunst CA, Li SH, Yi H, Mulroy JS, Kuemmerle S, Jones R, Rye D, Ferrante RJ, Hersch SM, Li XJ (1999) Nuclear and neuropil aggregates in Huntington's disease: relationship to neuropathology. J Neurosci 19:2522–2534

Hackam AS, Yassa AS, Singaraja R, Metzler M, Gutekunst CA, Gan L, Warby S, Wellington CL, Vaillancourt J, Chen N, Gervais FG, Raymond L, Nicholson DW, Hayden MR (2000) Huntingtin interacting protein 1 (HIP-1) induces apoptosis via a novel caspase-dependent death effector domain. J Biol Chem 275:41299–41308

Hedreen JC, Folstein SE (1995) Early loss of neostriatal striosome neurons in Huntington's disease. J Neuropathol Exp Neurol 54:105–120

Hodgson JG, Agopyan N, Gutekunst CA, Leavitt BR, LePiane F, Singaraja R, Smith DJ, Bissada N, McCutcheon K, Nasir J, Jamot L, Li XJ, Stevens ME, Rosemond E, Roder JC, Phillips AG, Rubin EM, Hersch SM, Hayden MR (1999) A YAC mouse model for Huntington's disease with full-length mutant hungtingtin, cytoplasmic toxicity, and selective striatal neurodegeneration. Neuron 23:181–192

Huang CC, Faber PW, Persichetti F, Mittal V, Vonsattel JP, MacDonald ME, Gusella JF (1998) Amyloid formation by mutant huntingtin: threshold, progressivity and recruitment of normal polyglutamine proteins. Somat Cell Mol Genet 24:217–233

Jackson GR, Salecker I, Dong X, Yao X, Arnheim N, Faber PW, MacDonald ME, Zipursky SL (1998) Polyglutamine-expanded human huntingtin transgenes induce degeneration of Drosophila photoreceptor neurons. Neuron 21:633–642

Kalchman MA, Graham RK, Xia G, Koide HB, Hodgson JG, Graham KC, Goldberg YP, Gietz RD, Pickart CM, Hayden MR (1996) Hungtingtin is ubiquitinated and interacts with a specific ubiquitin-conjugating enzyme. J Biol Chem 271:19385–19394

Kalchman MA, Koide HB, McCutcheon K, Graham RK, Nichol K, Nishiyama K, Kazemi-Esfarjani P, Lynn FC, Wellington C, Metzler M, Goldberg YP, Kanazawa I, Gietz RD, Hayden MR (1997) HIP1, a human homologue of S. cerevisiae Sla2p, interacts with membrane-associated huntingtin in the brain. Nat Genet 16:44–53

Kazantsev A, Preisinger E, Dranovsky A, Goldgaber D, Housman D (1999) Insoluble detergent-resistant aggregates form between pathological and nonpathological lengths of polyglutamine in mammalian cells. Proc Natl Acad Sci USA 96:11404–11409

Kazemi-Esfarjani P, Benzer S (2000) Genetic suppression of polyglutamine toxicity in Drosophila. Science 287:1837–1840

Kegel KB, Kim M, Sapp E, McIntyre C, Castano JG, Aronin N, DiFiglia M (2000) Huntingtin expression stimulates endosomal-lysosomal activity, endosome tubulation, and autophagy [In Process Citation]. J Neurosci 20:7268–7278

Kim M, Lee HS, LaForet G, McIntyre C, Martin EJ, Chang P, Kim TW, Williams M, Reddy PH, Tagle D, Boyce FM, Won L, Heller A, Aronin N, DiFiglia M (1999a) Mutant huntingtin expression in clonal striated cells: dissociation of inclusion formation and neuronal survival by caspase inhibition. J Neurosci 19:964–973

Kim M, Velier J, Chase K, Laforet G, Kalchman MA, Hayden MR, Won L, Heller A, Aronin N, DiFiglia M (1999b) Forskolin and dopamine D1 receptor activation increase huntingtin's association with endosomes in immortalized neuronal cells of striatal origin. Neuroscience 89:1159–1167

Klement IA, Skinner PJ, Kaytor MD, Yi H, Hersch SM, Clark HB, Zoghbi HY, Orr HT (1998) Ataxin-1 nuclear localization and aggregation: role in polyglutamine-induced disease in SCA1 transgenic mice. Cell 95:41–53

Kobayashi Y, Kume A, Li M, Doyu M, Hata M, Ohtsuka K, Sobue G (2000) Chaperones Hsp70 and Hsp40 suppress aggregate formation and apoptosis in cultured neuronal cells expressing truncated androgen receptor protein with expanded polyglutamine tract. J Biol Chem 275:8772–8778

Kouroku Y, Fujita E, Jimbo A, Mukasa T, Tsuru T, Momoi MY, Momoi T (2000) Localization of active form of caspase-8 in mouse L929 cells induced by TNF treatment and polyglutamine aggregates. Biochem Biophys Res Commun 270:972–977

Kuemmerle S, Gutekunst CA, Klein AM, Li XJ, Li SH, Beal MF, Hersch SM, Ferrante RJ (1999) Huntington aggregates may not predict neuronal death in Huntington's disease. Ann Neurol 46:842–849

Li H, Li SH, Johnston H, Shelbourne PF, Li XJ (2000) Amino-terminal fragments of mutant huntingtin show selective accumulation in striatal neurons and synaptic toxicity. Nat Genet 25:385–389

Li SH, Gutekunst CA, Hersch SM, Li XJ (1998) Interaction of huntingtin-associated protein with dynactin P150Glued. J Neurosci 18:1261–1269

Li XJ (2000) The early cellular pathology of Huntington's disease. Mol Neurobiol 20:111–124

Li XJ, Li SH, Sharp AH, Nucifora FC Jr, Schilling G, Lanahan A, Worley P, Snyder SH, Ross CA (1995) A huntingtin-associated protein enriched in brain with implications for pathology. Nature 378:398–402

Liu YF (1998) Expression of polyglutamine-expanded Huntingtin activates the SEK1-JNK pathway and induces apoptosis in a hippocampal neuronal cell line. J Biol Chem 273:28873–28877

Liu YF, Deth RC, Devys D (1997) SH3 domain-dependent association of huntingtin with epidermal growth factor receptor signaling complexes. J Biol Chem 272:8121–8124

Liu YF, Dorow D, Marshall J (2000) Activation of MLK2-mediated signaling cascades by polyglutamine-expanded huntingtin. J Biol Chem 275:19035–19040

Luthi-Carter R, Strand A, Peters NL, Solano SM, Hollingsworth ZR, Menon AS, Frey AS, Spektor BS, Penney EB, Schilling G, Ross CA, Borchelt DR, Tapscott SJ, Young AB, Cha JH, Olson JM (2000) Decreased expression of striatal signaling genes in a mouse model of Huntington's disease. Human Mol Genet 9:1259–1271

Mangiarini L, Sathasivam K, Seller M, Cozens B, Harper A, Hetherington C, Lawton M, Trottier Y, Lehrach H, Davies SW, Bates GP (1996) Exon 1 of the HD gene with an expanded CAG repeat is sufficient to cause a progressive neurological phenotype in transgenic mice. Cell 87:493–506

Martin EJ, Kim M, Velier J, Sapp E, Lee HS, Laforet G, Won L, Chase K, Bhide PG, Heller A, Aronin N, DiFiglia M (1999) Analysis of Huntingtin-associated protein 1 in mouse brain and immortalized striatal neurons. J Comp Neurol 403:421–430

Martin JB, Gusella JF (1986) Huntington's disease. Pathogenesis and management. New Engl J Med 315:1267–1276

Martindale D, Hackam A, Wieczorek A, Ellerby L, Wellington C, McCutcheon K, Singaraja R, Kazemi-Esfarjani P, Devon R, Kim SU, Bredesen DE, Tufaro F, Hayden MR (1998) Length of huntingtin and its polyglutamine tract influences localization and frequency of intracellular aggregates. Nat Genet 18:150–154

McCampbell A, Taylor JP, Taye AA, Robitschek J, Li M, Walcott J, Merry D, Chai Y, Paulson H, Sobue G, Fischbeck KH (2000) CREB-binding protein sequestration by expanded polyglutamine. Human Mol Genet 9:2197–2202

Miyashita T, Okamura-Oho Y, Mito Y, Nagafuchi S, Yamada M (1997) Dentatorubral pallidoluysian atrophy (DRPLA) protein is cleaved by caspase-3 during apoptosis. J Biol Chem 272:29238–29243

Miyashita T, Matsui J, Ohtsuka Y, Mami U, Fujishima S, Okamura-Oho Y, Inoue T, Yamada M (1999) Expression of extended polyglutamine sequentially activates initiator and effector caspases. Biochem Biophys Res Commun 257:724–730

Muchowski PJ, Schaffar G, Sittler A, Wanker EE, Hayer-Hartl MK, Hartl FU (2000) Hsp70 and hsp40 chaperones can inhibit self-assembly of polyglutamine proteins into amyloid-like firbrils. Proc Natl Acad Sci USA 97:7841–7846

Nasir J, Floresco SB, O'Kusky JR, Diewert VM, Richman JM, Zeisler J, Borowski A, Marth JD, Phillips AG, Hayden MR (1995) Targeted disruption of the Huntington's disease gene results in embryonic lethality and behavioral and morphological changes in heterozygotes. Cell 81:811–823

Ona VO, Li M, Vonsattel JP, Andrews LJ, Khan SQ, Chung WM, Frey AS, Menon AS, Li XJ, Stieg PE, Yuan J, Penney JB, Young AB, Cha JH, Friedlander RM (1999) Inhibition of caspase-1 slows disease progression in a mouse model of Huntington's disease. Nature 399:263–267

Portera-Cailliau C, Hedreen JC, Price DL, Koliatsos VE (1995) Evidence for apoptotic cell death in Huntington disease and excitotoxic animal models. J Neurosci 15:3775–3787

Reddy PH, Williams M, Charles V, Garrett L, Pike-Buchanan L, Whetsell WO Jr, Miller G, Tagle DA (1998) Behavioural abnormalities and selective neuronal loss in HD transgenic mice expressing mutated fill-length HD cDNA. Nat Genet 20:198–202

Reiner A, Albin RL, Anderson KD, D'Amato CJ, Penney JB, Young AB (1988) Differential loss of striatal projection neurons in Huntington's disease. Proc Natl Acad Sci USA 85:5733–5737

Rigamonti D, Bauer JH, De-Fraja C, Conti L, Sipione S, Sciorati C, Clementi E, Hackam A, Hayden MR, Li Y, Copper JK, Ross CA, Govoni S, Vincenz C, Cattaneo E (2000) Wild-type huntingtin protects form apoptosis upstream of caspase-3. J Neurosci 20:3705–3713

Ross CA (1997) Intranuclear neuronal inclusions: a common pathogenic mechanism for glutamine-repeat neurodegenerative diseases? Neuron 19:1147–1150

Sanchez I, Xu C-J, Juo P, Kakizaka A, Blenis J, Yuan J (1999) Caspase-8 is required for cell death induced by expanded polyglutamine repeats. Neuron 22:623–633

Sapp E, Ge P, Aizawa H, Bird E, Penney J, Young AB, Vonsattel JP, DiFiglia M (1995) Evidence for a preferential loss of enkephalin immunoreactivity in the external globus pallidus in low grade Huntington's disease using high resolution image analysis. Neurosci 64:397–404

Sapp E, Penney J, Young A, Aronin N, Vonsattel JP, DiFiglia M (1999) Axonal transport of N-terminal huntingtin suggests early pathology of corticostriatal projections in Huntington disease. J Neuropathol Exp Neurol 58:165–173

Satyal SH, Schmidt E, Kitagawa K, Sondheimer N, Lindquist S, Kramer JM, Morimoto RI (2000) Polyglutamine aggregates alter protein folding homeostosis in Caenorhabditis elegans. Proc Natl Acad Sic USA 97:5750–5755

Saudou F, Devys D, Trottier Y, Imbert G, Stoeckel ME, Brice A, Mandel JL (1996) Polyglutamine expansions and neurodegenerative diseases. In: Cold Spring Harbor Laboratory (eds) Cold Spring Habor Symposia Quantitative Biology 61, pp 639–647

Saudou F, Finkbeiner S, Devys D, Greenberg ME (1998) Huntingtin acts in the nucleus to induce apoptotsis but death does not correlate with the formation of intranuclear inclusions. Cell 95:55–66

Sawa A, Wiegand GW, Cooper J, Margolis RL, Sharp AH, Lawler JF Jr, Greenamyre JT, Synder SH, Ross CA (1999) Increased apoptosis of Huntington disease lymphoblasts associated with repeat length-dependent mitochondrial depolarization. Nat Med 5:1194–1198

Schilling G, Becher MW, Sharp AH, Jinnah HA, Duan K, Kotzuk JA, Slunt HH, Ratovitski T, Cooper JK, Jenkins NA, Copeland NG, Price DL, Ross CA, Borchelt DR (1999) Intranuclear inclusions and neuritic aggregates in transgenic mice expressing a mutant N-terminal fragment of huntingtin. Human Mol Genet 8:397–404

Sharp AH, Ross CA (1996) Neurobiology of Huntington's disease. Neurobiol Dis 3:3–15

Sharp AH, Loev SJ, Schilling G, Li SH, Li XJ, Bao J, Wagster MV, Kotzuk JA, Steiner JP, Lo A, Hedreen J, Sisodia S, Snyder SH, Dawson TM, Ryugo DK, Ross CA (1995) Widespread expression of Huntington's disease gene (IT15) protein product. Neuron 14:1065–1074

Shelbourne PF, Killeen N, Hevner RF, Johnston HM, Tecott L, Lewandoski M, Ennis M, Ramirez L, Li Z, Iannicola C, Littman DR, Myers RM (1999) A Huntington's disease CAG expansion at the murine Hdh locus is unstable and associated with behavioural abnormalities in mice. Human Mol Genet 8:763–774

Shimohata T, Nakajima T, Yamada M, Uchida C, Onodera O, Naruse S, Kimura T, Koide R, Nozaki K, Sano Y, Ishiguro H, Sakoe K, Ooshima T, Sato A, Ikeuchi T, Oyake M, Sato T, Aoyagi Y, Hozumi I, Nagatsu T, Takiyama Y, Nishizawa M, Goto J, Kanazawa I, Davidson I, Tanese N (2000) Expanded polyglutamine stetches interact with TAFII130, interfering with CREB-dependent transcription. Nat Genet 26:29–36

Sittler A, Walter S, Wedemeyer N, Hasenbank R, Scherzinger E, Eickhoff H, Bates GP, Lehrach H, Wanker EE (1998) SH3GL3 associates with the Huntingtin exon 1 protein and promotes the formation of polygln-containing protein aggregates. Mol Cell 2:427–436

Steffan JS, Kazantsev A, Spasic-Boskovic O, Greenwald M, Zhu YZ, Gohler H, Wanker EE, Bates GP, Housman DE, Thompson LM (2000) The Huntington's disease protein interacts with p53 und CREB-binding protein and represses transcription. Proc Natl Acad Si USA 97:6763–6768

Stenoien DL, Cummings CJ, Adams HP, Mancini MG, Patel K, DeMartino GN, Marcelli M, Weigel NL, Mancini MA (1999) Polyglutamine-expanded androgen receptors form aggregates that sequester heat shock proteins, proteasome components and SRC-1, and are suppressed by the HDJ-2 chaperone. Human Mol Genet 8:731–741

Thomas LB, Gates DJ, Richfield EK, O'Brien TF, Schweitzer JB, Steindler DA (1995) DNA end labeling (TUNEL) in Huntington's disease and other neuropathological conditions. Exp Neurol 133:265–272

Trottier Y, Devys D, Imbert G, Saudou F, An I, Lutz Y, Weber C, Agid Y, Hirsch EC, Mandel JL (1995a) Cellular localization of the Huntington's disease protein and discrimination of the normal and mutated form. Nat Genet 10:104–110

Trottier Y, Lutz Y, Stevanin G, Imbert G, Devys D, Cancel G, Saudou F, Weber C, David G, Tora L, Agid Y, Brice A, Mandel J-L (1995b) Polyglutamine expansion as a pathological epitope in Huntington's disease and four dominant cerebellar ataxias. Nature 378:403–406

Tukamoto T, Nukine N, Ide K, Kanazawa J (1997) Huntington's disease gene product, huntingtin, associates with microtubules in vitro. Brain Res Mol Brain Res 51:8–14

Turmaine M, Raza A, Mahal A, Mangiarini L, Bates GP, Davies SW (2000) Nonapoptotic neurodegeneration in a transgenic mouse model of Huntington's disease. Proc Natl Acad Sci USA 97:8093–8097

Velier J, Kim M, Schwarz C, Kim TW, Sapp E, Chase K, Aronin N, DiFiglia M (1998) Wild-type and mutant huntingtins function in vesicle trafficking in the secretory and endocytic pathways. Exp Neurol 152:34–40

Vonsattel JP, Myers RH, Stevens TJ, Ferrante RJ, Bird ED, Richardson EP Jr (1985) Neuropathological classification of Huntington's disease. J Neuropathol Exp Neurol 44:559–577

Wang GH, Mitsui K, Kotliarova S, Yamashita A, Nagao Y, Tokuhiro S, Iwatsubo T, Kanazawa I, Nukina N (1999) Caspase activation during apoptotic cell death induced by expanded polyglutamine in N2a cells. Neuroreport 10:2435–2438

Wanker EE, Rovira C, Scherzinger E, Hasenbank R, Walter S, Tait D, Colicelli J, Lehrach H (1997) HIP-I: a huntingtin interacting protein isolated by the yeast two-hybrid system. Human Mol Genet 6:487–495

Warrick JM, Paulson HL, Gray-Board GL, Bui QT, Fischbeck KH, Pittman RN, Bonini NM (1998) Expanded polyglutamine protein forms nuclear inclusions and causes neural degeneration in Drosophila. Cell 93:939–949

Wellington CL, Hayden MR (2000) Caspases and neurodegeneration: on the cutting edge of new therapeutic approaches. Clin Genet 57:1–10

Wellington CL, Ellerby LM, Hackam AS, Margolis RL, Trifiro MA, Singaraja R, McCutcheon K, Salvesen GS, Propp SS, Bromm M, Rowland KJ, Zhang T, Rasper D, Roy S, Thornberry N, Pinsky L, Kakizuka A, Ross CA, Nicholson DW, Bredesen DE, Hayden MR (1998) Caspase cleavage of gene products associated with triplet expansion disorders generates truncated fragments containing the polyglutamine tract. J Biol Chem 273:9158–9167

Wellington CL, Singaraja R, Ellerby L, Savill J, Roy S, Leavitt B, Cattaneo E, Hackam A, Sharp A, Thornberry N, Nicholson DW, Bredesen DE, Hyden MR (2000) Inhibiting caspase cleavage of huntingtin reduces toxicity and aggregate formation in neuronal and nonneuronal cells. J Biol Chem 275:19831–19838

White JK, Auerbach W, Duyao MP, Vonsattel JP, Gusella JF, Joyner AL, MacDonald ME (1997) Huntingtin is required for neurogenesis and is not impaired by the Huntington's disease CAG expansion. Nat Genet 17:404–410

Wyttenbach A, Carmichael J, Swartz J, Furlong RA, Narain Y, Rankin J, Rubinsztein DC (2000) Effects of heat shock, heat shock protein 40 (HDJ-2), and proteasome inhibition on protein aggregation in cellular models of Huntington's disease. Proc Natl Acad Sci USA 97:2898–2903

Yamamoto A, Lucas JJ, Hen R (2000) Reversal of neuropathology and motor dysfunction in a conditional model of Huntington's disease. Cell 101:57–66

Zeitlin S, Liu JP, Chapman DL, Papaioannou VE, Efstratiadis A (1995) Increased apoptosis and early embryonic lethality in mice nullizygous for the Huntington's disease gene homologue. Nat Genet 11:155–163

Subject Index

Printing (Computer to Film): Saladruck, Berlin
Binding: Stürtz AG, Würzburg